Relativity Contradictions Unveiled:

Kinematics, Gravity and Light Refraction

Robert J. Buenker

Apeiron
Montreal

Published by C. Roy Keys Inc.
4405, rue St-Dominique
Montreal, Quebec H2W 2B2 Canada
http://redshift.vif.com

First Published 2014

Library and Archives Canada Cataloguing in Publication

Buenker, Robert J., 1942-, author
Relativity contradictions unveiled : kinematics, gravity and light refraction / Robert J. Buenker.

Includes bibliographical references and index.

Issued in print and electronic formats.
ISBN 978-0-9920456-1-6 (pbk.).--ISBN 978-0-9920456-2-3 (ebook)

1. Special relativity (Physics). 2. Lorentz transformations. 3. Kinematics. 4. Gravity. 5. Light. 6. Refraction. 7. Contradiction. I. Title.

QC173.65.B83 2014 530.11 C2014-900677-2
C2014-901183-0

Front cover: Schematic diagram showing light rays emitted by stars to follow straight-line trajectories as they pass near the sun. Because of gravitational effects the speed of the light rays c′ is known to increase with gravitational potential, with the effect that the corresponding Huygens wave front gradually rotates away from the sun. As discussed in the text, the normal to a given wave front points out the direction from which the light appears to have come, causing the star images to be displaced by an angle Θ during solar eclipses.
Back cover: Angles of incidence Φ and refraction Φ′ as light enters water from free space. The corresponding angle Φ″ which the wave front makes with the interface upon entering a condensed medium can be determined by noting the direction from which the image of the light source appears to arrive inside the medium. Note that Φ″ is generally different than Φ′. As discussed in Chapter 9. C, measurement of all three angles enables the determination of both the normal refractive index n and the corresponding group refractive index n_g of the condensed medium.

Table of Contents

Dedication

To my dear wife, Marilyn Kay Lambrecht, for her continued support and advice over the past 44 years, and to our children, Michael, Timothy and Annie, for taking the time to listen to my ideas and giving me their always constructive criticism.

Acknowledgments

The author has many people to thank who have given me their encouragement and helpful suggestions over the years. Let me begin with Fr. Robert Dolter of Loras College, who fostered in me a healthy skepticism of traditional ideas in science in my formative years. I am especially grateful to Dipls. Phys. H.-P. Liebermann and Peter Funke for numerous helpful discussions about classical relativity theory and for their collaboration in developing computer programs to carry out the gravitational calculations using Schiff's method presented in the book. Special thanks are due to C. Roy Keys, my publisher, for his steady encouragement and professional advice, and to Dr. Thomas E. Phipps Jr. for his objective criticism of many of the topics of the book.

Robert J. Buenker
Feb. 11, 2014

Preface

The Global Positioning System (GPS) is recognized as one of the most useful applications of relativity theory. It relies in the main on two results of Einstein's special theory of relativity (STR) introduced in his seminal paper of 1905. Ever since then it has been hailed by the scientific community as a completely reliable guide into the realm of high-energy physics. Nonetheless, closer examination of the underlying theory of the GPS methodology reveals some clear inconsistencies with STR. For one thing, satellites are not inertial systems since they are subject to constant acceleration as they make their way around the earth. By contrast, it has been a longstanding belief among physicists that the range of applicability of STR is confined exclusively to systems that are moving in pure translation and are thus not subject to any external force. One of the consequences of this restriction is Einstein's Symmetry Principle (ESP), which states that observers in relative motion will each find that it is the other's clock that has been slowed by the effects of time dilation. Yet the GPS methodology assumes that the clocks carried on satellites run objectively slower than identical counterparts on the earth's surface (after corrections for gravitational effects on clock rates have been taken into account). The asymmetric adjustment of clock rates on the satellites to make them run at the same rate as those on the ground is well known to lead to accurate timing results. In their landmark experiment with atomic clocks carried on board circumnavigating airplanes, which is the basis for GPS, Hafele and Keating tried to justify the application of STR to estimate clock rate variations by pointing out that the non-rotating axes of the earth did serve as at least an approximately inertial system.

However, a more critical assessment is that the asymmetry in clock rates demonstrates that something is fundamentally wrong with STR itself, *i.e.* something that makes it inapplicable to situations in which even the slightest force acts on the objects under consideration. The theory itself is based on the Lorentz space-time transformation (LT) which Einstein and Lorentz and their contemporaries had developed as an alternative to the classical Galilean transformation (GT). The latter had proved incapable of explaining the results of fundamental research such as the Michelson-Morley experiment carried out in the late 19th century. Two of the most famous predictions of the LT are Fitzgerald-Lorentz length contraction (FLC) and time dilation. On this basis there are *two* quite distinct ways to measure the dimensions of an object such as a metal rod located on a satellite. On the one hand, it is claimed on the basis of the FLC that the length L of the rod oriented perpendicularly to the direction of motion is the same for a stationary observer O on the earth's surface as the corresponding value L′ measured by his counterpart O′ co-moving with the rod, *i.e.* $L = L'$. The GPS technology does not rely on the FLC to measure distances, however. Instead, it uses another method which is based on Einstein's second postulate of relativity: the elapsed time for light to pass between two end points is simply multiplied with the speed of light c to determine the distance traveled. According to STR and the LT, however, clocks run slower on the accelerated object (time dilation) so that the elapsed time ΔT measured by O is greater than the corresponding value $\Delta T'$ measured by O′. Since both observers must agree on the speed of light according to Einstein's second postulate, it follows that $L = c\Delta T$ must be greater than $L' = c\Delta T'$, *i.e.* $L > L'$ (length expansion as opposed to length contraction deduced from the FLC). There is thus a clear contradiction between the results of Einstein's theory depending on whether the FLC is used or the light-speed postulate. That is a totally unacceptable state of affairs. Any theory that gives opposite predictions for the same event is deficient and therefore needs to be modified in such a way as to remove the contradiction. In short, the LT, despite its long acceptance by physicists, is quite simply wrong.

The amazing thing about the above proof [which is referred to as the "clock riddle;" see R. J. Buenker, *Apeiron* **19**, 84 (2012)] is that this problem with the LT was not detected much earlier. No experiments whatsoever are required to demonstrate it. It is merely a matter of applying the various claims of the theory in a straightforward manner to the same problem. The question thus arises as to which of the two opposing predictions of Einstein's STR is actually correct. Do lengths expand when clocks slow down or do they stay the same or even contract as the FLC would have us believe? The answer was provided in 1938 by Ives and Stilwell with their investigation of the transverse Doppler effect. They found that the wavelength of light emitted from an accelerated source *increases* after correction for the conventional first-order Doppler effect is made. They inferred from their result that time dilation had occurred in the rest frame of the light source since the increase in wavelength is only consistent with a decrease in the corresponding frequency by virtue of Einstein's light-speed constancy postulate. Overlooked in this analysis is the deduction based on the first postulate, Galileo's Relativity Principle (RP): the dimensions of all objects in the rest frame of the light source must have increased in the same proportion as the wavelength since otherwise an observer there would be able to detect a change in his state of uniform motion on this basis. The conclusion is that length expansion accompanies time dilation, thereby contradicting the FLC; moreover, the amount of the expansion is the same in all directions (isotropic length expansion). In the ensuing years, many different experiments have been carried out that confirm this interpretation.

How did Einstein make the error that led to the LT and FLC? And how can the experimental result of length expansion accompanying time dilation be incorporated into relativity theory without eliminating the many confirmed predictions of STR in other applications. Fortunately, the answer to both questions can be found in Einstein's 1905 paper. He made an undeclared assumption about a normalization condition in the derivation of the LT that is directly responsible for his error [R. J. Buenker, *Phys. Essays* **26**, 494 (2013)]. One only has to replace this assumption with another which is based on solid evidence derived from the GPS methodology and earlier experiments with atomic clocks reported by Hafele and Keating in 1971, namely that the rates of clocks in uniform relative motion are strictly proportional to one another: $\Delta T' = \Delta T/Q$. The result is an alternative Lorentz transformation (ALT) that also satisfies Einstein's two postulates and is completely consistent with the relativistic velocity transformation (VT) of STR. All of the confirmed space-time predictions of the conventional theory can be derived exclusively from the VT, so the latter characteristic of the ALT is of utmost importance in the present discussion. At the same time, it eliminates the contradiction in the LT and brings relativity theory into agreement with experiment by replacing the FLC with isotropic length expansion.

More generally, the ALT leads to a perfectly objective theory of measurement. It is consistent with the asymmetric timing relationships assumed in the GPS methodology, unlike what is predicted from Einstein's ESP. The proportionality constant Q is seen to be a conversion factor between the respective units of time in different rest frames. A key goal of the resulting theory is to enable the computation of this constant on the basis of a minimum of information. A review of available experimental data shows that analogous proportionality relationships hold for all physical properties. The corresponding conversion factors are seen to be integral powers of Q in each case.

The resulting theory of measurement is therefore characterized by a *uniform scaling* principle. The laws of physics are still the same in all inertial systems but the units on which they are based vary in a well-defined manner between them. This constitutes an amended version of Galileo's Relativity Principle. Moreover, it is shown that another scaling parameter exists for gravitational relationships. Its value, referred to as S, can be deduced from Einstein's Equivalence Principle (EP), also with a minimum of information. Again, the conversion factors for other physical properties are seen to be integral powers of this quantity, but the values of these integers are shown to be different in most cases than for the kinetic scaling. Following earlier work [L. I. Schiff, *Am. J. Phys.* **28**, 340 (1960)], it is found that the combination of gravitational and kinetic scaling of units allows one to compute key quantities such as the angle of displacement of star images during solar eclipses and the angle of precession of Mercury's perihelion. There is quantitative agreement between the results of this simplified theory and those of Einstein's General Theory of Relativity (GTR). The concept of curved space in the latter theory is replaced by a special scaling of radial velocities to be used by an observer in a hypothetical gravity-free region of space. Perhaps most interestingly, it is shown that the displacement angle of star images can be computed by assuming that individual light rays travel in perfectly straight lines. Use of Huygens' principle indicates that changes in the speed of light with distance from the sun lead to a rotation of the light waves that gives the illusion that the stars have moved from their normal locations in the absence of solar eclipses.

The uniform scaling principle is directly applicable to GPS methodology. As discussed in previous work [R. J. Buenker,

G. Golubkov, M. Golubkov, I. Karpov and M. Manzheliy, "Relativity Laws for the Variation of Rates of Clocks Moving in Free Space and GPS Positioning Errors Caused by Space-weather Events," in *Global Navigation Systems* (ISBN 980-307-843-9; InTech-China, 2013)], the values of the kinetic and gravitational scaling parameters described above can be computed on the basis of the satellite's current location and speed along its trajectory at any given time. This information is sufficient to enable the instantaneous calibration of onboard clocks on satellites to ensure that they run at exactly the same rate as their counterparts on the earth's surface with which they are communicating. This procedure avoids any physical adjustment of clock rates and is applicable to any orbit over indefinite periods of time, starting from the moment of the GPS satellite's launch. One merely needs to multiply the readings of unadjusted (proper) clocks on the satellites with the appropriate factors determined on the basis of the uniform scaling principle.

As is often true in general, removing an error in a physical theory leads to many simplifications, principally the concept of uniform scaling of physical units in the present case. It is much easier to comprehend a theory of measurement that is completely objective instead of believing that two clocks can each be running slower than the other or that two masses are each heavier than the other. How the appropriate conversion factors can be calculated between the natural physical units employed by two different observers is the key to relating their respective experimental results for the same quantity, and this is the overriding subject of the present book.

About the Author

Robert J. Buenker was born in Dubuque, Iowa on May 6, 1942 to Joseph F. and Melita E. Buenker (nee Ferring). He received his undergraduate training at Loras College (BS in chemistry and mathematics, 1963) and did his graduate work at Princeton University (PhD in physical chemistry, 1966). After a postdoctoral year at Michigan State University, he joined the chemistry faculty at the University of Nebraska-Lincoln in 1967, where he attained the rank of Full Professor in 1973. He moved briefly to the University of Bonn, Germany in 1976 and then on to the Bergische Universität Wuppertal in 1977, where he became emeritus Professor of Chemistry in 2007. His main area of research is in the field of molecular physics. He has developed a series of computer programs that are in wide use throughout the world to carry out quantum mechanical calculations of the electronic structure of small molecules. He has had leaves of absence in Russia, Japan, China, Greece and the United States to pursue this research. The main applications are in molecular spectroscopy, collision dynamics, resonance states and positronic molecules. Some of this work involves relativistic effects in molecules containing heavy atoms, and this has led to his more recent interest in classical relativity theory, which is the subject of the present book. He has over 500 publications in refereed journals in both chemistry and physics.

Chapter 1

Introduction

One of the most basic precepts of relativity theory dating back to Galileo is that it is impossible for an observer to distinguish one inertial frame from another based on *in situ* measurements alone. If one measures a light frequency *in situ* for a given source, for example, the same value is always obtained independent of the state of motion of the laboratory. This is true even though it is known from experiment that the value of the frequency changes with the speed of the source when measured by an observer who is in relative motion to it. The reason the *in situ* value is constant is that the rates of all clocks change *in the same proportion* as the speed of the source is varied. The same argument (the relativity principle or RP) applied to the measurement of lengths gives a similar result, that *in situ* measurements also are incapable of detecting any change in the distance between two fixed points in a given inertial system.

Galileo's motivation for introducing the RP seems to have been closely tied up with his astronomical observations. His controversial heliocentric view of the earth's surroundings led him to conclude that we are in orbit around the sun with an almost unfathomable speed of 30 kms^{-1}. This state of affairs had gone unnoticed for centuries and there had to be a fundamental explanation for this lack of perception of man's true circumstances. When Galileo gave his interpretation of the RP in 1632, he used the example of a ship on the open seas as an illustration. He carefully distinguished between the case of the ship under the influence of wind and wave and that in which it calmly floats over the surface of the water unperturbed by any external forces. His main point was that in the latter situation observers shut away in the hold of the ship that were unable to view the outside world would find it impossible to detect whether the ship was moving relative to the shore or lay at rest in the harbor. He allowed the passengers perfect latitude to conduct any and all experiments of their choosing, including the observation of insects flying from one end of the ship to the other and fish swimming in a bowl of water.[1]

The point that will receive special emphasis in the present work is that the RP opens up the real possibility that we are in a constant state of change as we move about the universe. The fictional example of *Alice in Wonderland* is at the heart of this speculation. She notices that her surroundings have changed dramatically because of something that has suddenly occurred. The question in real life is how to quantify these changes for specific physical properties. The lesson taught by Galileo is not that everything is the same for the ship's passengers regardless of their state of constant motion, but rather that there might be great distinctions that simply go unnoticed because of the lack of a definitive means to recognize them on an experimental basis. "Everything is relative" is the watchword. Our physical dimensions might change enormously as we go from one state of constant motion to the other, but if everything else in our surroundings changes in exactly the same proportion, it will be impossible to notice any difference.

Changing one's state of motion means that there has been acceleration. Newton concluded with his Second Law of Kinematics that this can only occur if there is an applied force, which in turn implies that any distinctions in the physical properties of the ship's passengers come about because of this force. Theoretically, such changes might be detectable inside Galileo's ship during such an acceleration phase. For example, one might notice that the rates of physical processes are slowing down or that the masses of objects in the frame of rest are increasing during the actual period when the force is being applied. Another means of changing the state of motion might well be expected to have analogous (albeit different) effects, for example, when gravity is the cause.

A much more obvious characteristic of motion is the fact that different observers can disagree as to how fast a given object moves relative to them. A set of equations was developed to provide a quantitative relationship between the measured velocities of two observers who were themselves in relative motion to one another. The origin of these equations is unclear but it became known as the Galilean velocity transformation (GVT) over time. It became virtually self-evident to solve the problem using vector addition to obtain the three velocity components for one observer on the basis of the corresponding values of the other. It was also assumed that the clocks of the two observers always run at the same rate ($dt = dt'$). Conceptually, space and time were considered to be completely distinct quantities. Multiplication of the three individual velocity components for both observers with an elapsed time then leads to the Galilean transformation (GT) of space and time coordinates. At any one time on this basis one could plot the positions of a given object in a three-dimensional Cartesian space centered at an origin that could be arbitrarily chosen for each observer.

Experimental developments eventually caused scientists to reassess the validity of the GT. In particular, Maxwell's theory of electricity and magnetism published in 1864 posed a seemingly irreconcilable problem for the classical velocity transformation. On the one hand, because of the RP Maxwell's theory should be equally applicable in all inertial systems, while on the other, it seemingly required that the speed of light c be the same in each of them. The latter requirement was in conflict with the additivity principle of the GVT: if two inertial systems have a relative speed v, the seemingly incongruous relation, $c + v = c$, must hold for them. This state of affairs led to much speculation about an "aether" that somehow interacted with the electromagnetic field accompanying the light in such a way as to maintain the required constancy of the speed of light.

An alternative approach to dealing with the above uncertainty was to develop a new space-time transformation, which unlike the Galilean, would leave Maxwell's equations invariant. The first author to come up with such a solution was Voigt in 1887.[2] It soon became clear that there was more than one transformation that satisfied this goal, however. At about the same time, Michelson and Morley performed a new experiment[3] which gave strong support to the light-speed constancy concept. There followed an intense period of discussion about how to reconcile the theory of electricity and magnetism with the RP, led by Lorentz, Larmor, Poincaré, FitzGerald and others.[4]

This activity culminated with the publishing of Einstein's landmark 1905 paper[5] in which he introduced what has come to be known as the Special Theory of Relativity (STR). He based his theoretical approach on two postulates, a version of Galileo's RP and the constancy of light in free space, but he also made the decisive step to eliminate the need of an aether to explain the new experimental results. The centerpiece of his new theory is the Lorentz transformation (LT), which differs from Voigt's by a common factor in each of the four space-time equations. The same version of the transformation was given earlier by Larmor, Lorentz and Poincaré.[4] It has many distinctive consequences, and their detailed consideration will form a large part of the following discussion. Among these are Lorentz invariance, time dilation, FitzGerald-Lorentz length contraction (FLC), inextricable space-time mixing, a subjective view of the measurement process that allows observers to disagree as to which clock runs slower or which distance is longer, remote non-simultaneity of events, and at least the theoretical possibility of time reversal, just to name some of the most well-known ones. The LT also leads directly to the relativistic velocity transformation (VT) that ensures satisfaction of the light-speed postulate as well as providing straightforward predictions of such well-known empirical observations as the aberration of starlight at the zenith and the Fresnel light-drag experiment. The original derivation of the LT that Einstein gave in his 1905 paper is therefore the obvious starting point for the next part of this book.

Chapter 2

The Lorentz Transformation

The purpose of any transformation of space and time coordinates is to express the relationship between the measured positions of a given object (or event) made by two observers who are in relative motion. To simplify matters it is assumed that both the observers and object are in uniform translation, which means that they are not subject to any outside force. Their rest frames are referred to as inertial systems for this reason. In anticipation of the subsequent discussion of gravitational effects, it is further assumed that both observers and the object are always at the *same gravitational potential,* for example, at an infinite distance from any massive bodies.

A. Galilean Transformation as Precursor

Classical physicists assumed that there can be a separate coordinate system for each rest frame. In general, the coordinates of the object of the measurement are given in terms of the four numbers for the three spatial coordinates and the single time variable: x, y, z and t for one observer O in inertial system S, and x′, y′, z′ and t′ for the other (O′ in S′). One can assume without loss of generality that the direction of relative motion of S and S′ is along their mutual x-axis. Further, it is assumed that the coordinate axes coincide at some initial time ($t = t' = 0$) and indeed that $t = t'$ for all future measurements. The latter is a key assumption in the classical theory that will be removed subsequently when attention turns to developing a suitably relativistic transformation. This means that the x, y, z coordinates for the object as measured by O are exactly the same at $t = t' = 0$ as the corresponding x′, y′, z′ values measured by O′. However, they start to differ as the time of observation proceeds. A further assumption is made in this respect, namely that $y = y'$ and $z = z'$ at all times. To the classical physicist it seemed obvious that O and O′ would also agree on the values of these coordinates because their relative motion is perpendicular to both the y and z axes. As with the $t = t'$ assumption, it will prove advisable to be skeptical of this conclusion in the general case, particularly when very high speeds are involved.

The only differences in the measured values of the two observers therefore occur for their respective x coordinates. Specifically, it is assumed that if O′ moves with speed v in the positive x direction relative to O, then $x' = x - vt$. The sign convention is important in this equation. If O′ moves in the positive x direction relative to O, then $v > 0$; otherwise, $v < 0$. The resulting full transformation of the respective coordinates is therefore:

$$\begin{aligned} x' &= x - vt \\ y' &= y \\ z' &= z \\ t' &= t. \end{aligned} \tag{1}$$

It will be referred to as the Galilean transformation (GT) in the following discussion.

There is a companion transformation of the velocity components $u_x = dx/dt$, $u_x' = dx'/dt'$, *etc.*, of the object. It is obtained from eq. (1) by differentiation of the variables in each spatial equation with respect to $t = t'$:

$$\begin{aligned} u_x' &= u_x - v \\ u_y' &= u_y \\ u_z' &= u_z. \end{aligned} \tag{2}$$

These equations will subsequently be referred to as the Galilean velocity transformation (GVT). The respective components in the two perpendicular directions are simply assumed to be equal, the obvious consequence of the previous assumption about the y, y′ and z, z′ coordinates in arriving at the GT.

A key aspect of the GT is illustrated in Fig. 1. The position of an observer O is fixed in the diagram, *i.e.* he has the same position at $t = 0$ and later at $t = dt$. Observer O′ is moving away from O with constant speed v. He is at position M_1 at $t = 0$ and M_2 at

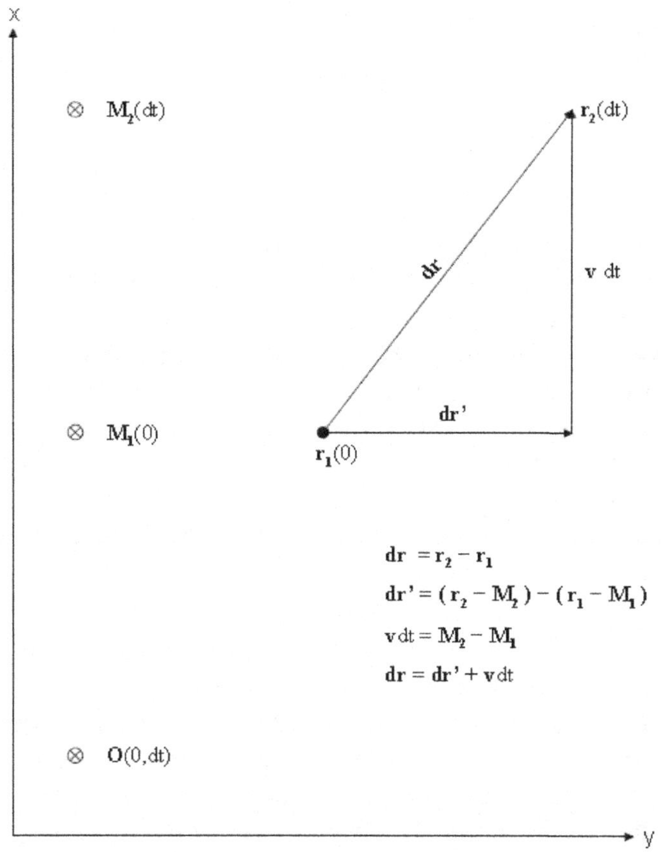

Fig. 1. Diagram illustrating the derivation of the Galilean transformation (GT). An object moves between positions $\mathbf{r}_1$ and $\mathbf{r}_2$ in elapsed time dt at the same time that observer M travels from $\mathbf{M}_1$ to $\mathbf{M}_2$ with speed v relative to a fixed origin **O**. The change in the location of the object relative to M is denoted as **dr'**, which is related by the GT to the corresponding change **dr** relative to O. Note that the same unit of length is employed throughout.

$t = dt$. Both observers are watching the progress of an object as it moves diagonally across the diagram. It has position $\mathbf{r}_1$ at $t = 0$ and $\mathbf{r}_2$ at $t = dt$. All of the above coordinates can also be described in terms of three-dimensional position vectors defined relative to the constant position of O. The vector distance travelled by the object is then given as $\mathbf{dr} = \mathbf{r}_2 - \mathbf{r}_1$ from the perspective of observer O. The corresponding quantity measured by O′ is $\mathbf{dr}' = \mathbf{r}_2' - \mathbf{r}_1'$. At the same time, the vector distance travelled by O′ relative to O's constant position is given as $\mathbf{M}_2 - \mathbf{M}_1 = \mathbf{v}dt$, where $\mathbf{v}$ is the corresponding velocity vector describing the relative motion of the two observers.

Combining the various vectors algebraically leads to the velocity addition theorem or GVT:

$$\mathbf{dr} = \mathbf{dr}' + \mathbf{v}dt. \tag{3}$$

One of the most important aspects about the above derivation is that it does not require a definite coordinate system to obtain the result. This is in stark contrast to the way the GT itself has been derived, in which case it was necessary to specify that O′ is moving along the common x-axis of the two coordinate systems. The choice of the x-axis for this purpose is clearly arbitrary, so from a conceptual standpoint this vector approach has clear advantages. Differentiation of eq. (3) leads to the corresponding vector equation for the GVT, with the definitions $\mathbf{u} = \mathbf{dr}/dt$ and $\mathbf{u}' = \mathbf{dr}'/dt$ (note that one could just as well use dt′ in both cases or dt′ in the second to make things look more symmetric):

$$\mathbf{u} = \mathbf{u}' + \mathbf{v}. \tag{4}$$

It will be shown subsequently that the corresponding relativistic velocity transformation (VT) cannot be expressed in such a simple vector form, so it is important to see that both eqs. (3-4), as well as eqs. (1-2), are based on assumptions that are tentative at best, even though they have a pleasingly intuitive character about them. It should be noted that eqs. (3-4) involve exclusively vector differences. That is the reason why it is completely arbitrary how one chooses the specific origin of the coordinate systems used by O and O′ in deriving these relationships. That point will also come up again in the ensuing discussion of suitably relativistic space-time transformations.

B. Derivation of the Lorentz Transformation

As discussed in the Introduction, the GT first became seriously questioned because of experimental results obtained in the latter half of the 19th century. The Michelson-Morley experiment seems[3] to have been decisive in this respect, but Fizeau's successful demonstration of the Fresnel light-drag phenomenon and Maxwell's derivation of the equations of electricity and magnetism were also indicative that something was fundamentally wrong with the classical theory of motion. Emphasis began to be placed on developing a different space-time transformation that a) left

Maxwell's equations invariant and b) ensured that the speed of light in free space is exactly the same for all observers and light sources regardless of their own state of motion. Einstein used the latter condition as the second postulate of relativity in his 1905 paper.[5]

As with the GT, it was assumed that both observers and the mutual object of measurement are in uniform translation, from which it could safely be assumed that the required transformation was linear, *i.e.* the coefficients a_{ij} in the four-dimensional space-time transformation are all constant. It is often assumed that this condition implies that each of these coefficients can only be a function of the relative speed v of the two inertial systems S and S′, but that conclusion is not justified on the basis of the linearity condition of itself. One simply writes down the most general set of such linear equations as follows:[6]

$$\begin{aligned} t' &= a_{00}t + a_{01}x + a_{02}y + a_{03}\,z \\ x' &= a_{10}t + a_{11}x + a_{12}y + a_{13}\,z \\ y' &= a_{20}t + a_{21}x + a_{22}y + a_{23}\,z \\ z' &= a_{30}t + a_{31}x + a_{32}y + a_{33}\,z. \end{aligned} \tag{5}$$

It is also assumed, as before with the GT, that the two sets of coordinate axes coincide when $t = t' = 0$. This allows one to assume on the basis of the homogeneity and isotropy of inertial frames that a number of a_{ij} values must be zero. For example, a point on the y axis has $x = 0$ and $z = 0$ at $t = 0$, so the above coincidence condition requires that $x' = z' = 0$ as well. This condition can only be fulfilled if $a_{12} = a_{32} = 0$. Similar results hold for the other four off-diagonal coefficients in the space-space part of the matrix.

At this point in the derivation it is useful to consider differences of coordinates (intervals) for two separate events, *i.e.* $\Delta t = t_2 - t_1$ *etc.* The corresponding equations thus become:

$$\begin{aligned} \Delta t' &= a_{00}\Delta t + a_{01}\Delta x + a_{02}\Delta y + a_{03}\Delta z \\ \Delta x' &= a_{10}\Delta t + a_{11}\Delta x \\ \Delta y' &= a_{20}\Delta t + a_{22}\Delta y \\ \Delta z' &= a_{30}\Delta t + a_{33}\Delta z. \end{aligned} \tag{6}$$

Note that this decision to work with intervals removes the need to assign specific origins for the two coordinate systems, but rather only to assume that the relative velocity of S and S′ is along their respective x, x′-axes. The next step is to consider successive locations of the object when it is also moving with speed v along the x, x′ axes.

Consequently, the intervals in S satisfy $\Delta x = v\Delta t$, $\Delta y = \Delta z = 0$, whereas $\Delta x' = \Delta y' = \Delta z' = 0$ in S′. Substitution in eq. (6) shows that $a_{10} + va_{11} = 0$, $a_{20} = a_{30} = 0$. Reversing this relationship so that the object is at rest in S leads to the corresponding set of results: $a_{10} + va_{00} = 0$, $a_{20} = a_{30} = 0$. It is thus clear that $a_{11} = a_{00}$ and $a_{10} = -v\, a_{00}$, whereupon the relations in eq. (6) simplify to:

$$\begin{aligned} \Delta t' &= a_{00}\Delta t + a_{01}\Delta x + a_{02}\Delta y + a_{03}\Delta z \\ \Delta x' &= -va_{00}\Delta t + a_{00}\Delta x \\ \Delta y' &= a_{22}\Delta y \\ \Delta z' &= a_{33}\Delta z. \end{aligned} \tag{7}$$

The conditions of homogeneity and isotropy of space can be used further by considering events that are simultaneous in either of the rest frames. For example, if the events have the same x coordinate ($\Delta x = 0$) and $\Delta t = 0$, it follows that $\Delta x' = 0$ in S′. The time of the event cannot depend on its distance from the x (or x′) axis, so the two events must also be simultaneous in S′ ($\Delta t' = 0$). Therefore, from the first relation of eq. (7) it follows for such events in either the (x,y) or (x,z) planes, *i.e.*, with either $\Delta y \neq 0$ or $\Delta z \neq 0$, that both the coefficients a_{02} and a_{03} must vanish. Finally, isotropy also demands that there not be an *azimuthal twist* that causes the proportions of the y, z and y′, z′ coordinates to differ, from which it can be concluded that $a_{22} = a_{33}$. As a result the transformation in eq. (7) is further simplified to:

$$\begin{aligned} \Delta t' &= a_{00}\Delta t + a_{01}\Delta x \\ \Delta x' &= -va_{00}\Delta t + a_{00}\Delta x \\ \Delta y' &= a_{22}\Delta y \\ \Delta z' &= a_{22}\Delta z. \end{aligned} \tag{8}$$

The above equations still contain three undefined coefficients: a_{00}, a_{01} and a_{22}. In order to specify them further, it is necessary to impose the condition that the speed of a light pulse in free space be the same ($c = 2.99792458 \times 10^8\ ms^{-1}$) for both O and O′ independent of their respective locations. It is quite important to recognize, however, that information about velocities can never be sufficient in itself to specify the space-time transformation to more than within a common factor in all four equations. This is because velocities are ratios of space and time coordinates and thus the above factor will simply cancel upon performing the necessary division to obtain them. This fact can be made more transparent by defining the transformation in terms of the alternative constants: $\varepsilon = a_{22}$, $A = a_{00}/a_{22}$ and $B = a_{01}/\ a_{00}$. Substitution in eq. (8) then gives:

$$
\begin{aligned}
\Delta t' &= \varepsilon A(\Delta t + B\Delta x) \\
\Delta x' &= \varepsilon A\ (\Delta x - v\Delta t) \\
\Delta y' &= \varepsilon \Delta y \\
\Delta z' &= \varepsilon \Delta z.
\end{aligned}
\qquad (9)
$$

The constant B can then be evaluated by forming the ratio $\Delta x'/\Delta t'$ from eq. (9). Since B is a constant and the transformation is to be perfectly general, it is permissible to evaluate it for a particular case, specifically for a light pulse moving along the common x, x′-axis of the two coordinate systems and imposing the condition that $\Delta x/\Delta t = c$ as well. Thus,

$$\Delta x'/\Delta t' = c = (\Delta x - v\Delta t)/(\Delta t + B\Delta x) = (c - v)/(1 + Bc), \qquad (10)$$

from which one concludes that $B = -v/c^2$.

Evaluating the constant A is only slightly more complicated. First of all, since A is a constant, one is free to choose any value for ε in the evaluation. The simplest choice is $\varepsilon = 1$. A light pulse satisfies the general condition given below because its speed is equal to c when measured in both S and S′:

$$\Delta x'^2 + \Delta y'^2 + \Delta z'^2 - c^2\Delta t'^2 = \Delta x^2 + \Delta y^2 + \Delta z^2 - c^2\Delta t^2 = 0. \qquad (11)$$

Since $\varepsilon = 1$, $\Delta y = \Delta y'$ and $\Delta z = \Delta z'$ in eq. (9), and thus the corresponding squared terms in eq. (11) simply cancel. Further substitution from eq. (9) with $B = -v/c^2$ then gives:

$$
\begin{aligned}
\Delta x'^2 - c^2\Delta t'^2 &= A^2(\Delta x^2 - 2v\Delta x\Delta t + v^2\Delta t^2) \\
&- c^2A^2(\Delta t^2 - 2vc^{-2}\Delta x\Delta t + v^2c^{-4}\Delta x^2) = \Delta x^2 - c^2\Delta t^2
\end{aligned}
\qquad (12)
$$

Since this equation must hold for all values of Δx and Δt, one can equate the individual terms in eq. (12) for Δx^2, with the result:

$$A^2(1 - v^2c^{-2}) = 1. \qquad (13)$$

The conclusion is thus

$$A = \gamma(v) = (1 - v^2c^{-2})^{-0.5}, \qquad (14)$$

whereby the constant A is replaced by its standard designation γ to be used in the ensuing discussion [note that the coefficients of $\Delta x\Delta t$ and Δt^2 are also equal in eq. (12) with this choice of the value for A]. Substitution in eq. (9) then leads to the following general form for the Lorentz transformation (GLT), given in terms of intervals rather than the corresponding coordinates themselves:

$$\begin{aligned} \Delta t' &= \varepsilon\gamma(\Delta t - vc^{-2}\Delta x) = \varepsilon\gamma\eta^{-1}\Delta t \\ \Delta x' &= \varepsilon\gamma(\Delta x - v\Delta t) \\ \Delta y' &= \varepsilon\Delta y \qquad (15) \\ \Delta z' &= \varepsilon\Delta z\,, \end{aligned}$$

with $\eta = (1 - vc^{-2}\Delta x/\Delta t)^{-1} = (1 - vu_xc^{-2})^{-1}$, *i.e.* with $u_x = \Delta x/\Delta t$.

The GLT was first published in this form by Lorentz in 1899.[4,7] It needs to be emphasized that, despite the fact that a specific value for ε has been used at various stages of the above derivation, no definitive assignment can be made for it based on the light-speed postulate alone. There is a degree of freedom in the GLT, expressed through the common (normalization) factor ε, which simply goes undefined in the derivation. Einstein was clearly aware of this point (see p. 900 of Ref. 5 after the transformation equations; he used ϕ in his notation instead of Lorentz's ε). He nonetheless goes on to conclude that $\varepsilon = 1$ is the only acceptable value for this quantity. The same conclusion was reached earlier by both Lorentz and Poincaré[4] and thus it is interesting to consider the justifications that have heretofore been given for this choice.

C. Determination of the Normalization Factor

Einstein[5] states on p. 900 of Ref.[5] that "ϕ is an as yet undetermined function of v," that is, of the relative speed of the two inertial systems involved in the space-time transformation. It has been pointed out in the previous section that the required linearity of these equations merely demands that all the a_{ij} quantities in eq. (5) be constants. Thus, Einstein's assertion that the normalization function can only depend on v, which is assumed to be constant in the overall derivation, is more specific than need be. For example, the quantity η in the GLT of eq. (15) is a function of the component of the object's speed in the parallel direction, $u_x = \Delta x/\Delta t$, which is also a constant under the conditions of the derivation.

Nonetheless, Einstein[5] continued by noting that *under these circumstances* the product of the transformation for a given value of v with its inverse requires that

$$\phi(v)\,\phi(-v) = 1. \tag{16}$$

At the same time, he noted that symmetry requires that the effect of the transformations in the perpendicular directions must be the same, which therefore implies that ϕ (v) is an even function. Thus, $\phi = 1$ is the only possible solution that is consistent with the fact that ϕ (0) = 1. On this basis he concluded that the only space-time transformation that is consistent with the light-speed constancy condition as well as the homogeneity and isotropy properties of space is that which is obtained by setting $\phi = \varepsilon = 1$ in the GLT, namely:

$$\begin{aligned} t' &= \gamma(t - vc^{-2}x) = \gamma\eta^{-1}t \\ x' &= \gamma(x - vt) \\ y' &= y \\ z' &= z. \end{aligned} \tag{17}$$

It is this set of space-time relations that has come to be known as the Lorentz transformation (LT). As mentioned previously, the same equations had been given earlier by Larmor, Lorentz and Poincaré.[4] In his work,[8] Poincaré gave a different justification for using $\varepsilon = 1$ in the LT (he was the first to apply the name "Lorentz transformation" to these equations[4]). He pointed out that successive transformations satisfied the postulates of a group, and he argued that this property was essential for such a fundamental relationship. Einstein[4,5] also made a similar remark in his paper. Nonetheless, this is also not a compelling argument, and thus still leaves open the possibility that another choice of ε might actually be required. This assessment receives additional support from the fact that other authors have sought to justify the $\varepsilon/\phi = 1$ choice for the LT in different ways than Einstein. For example, Goldstein[9] in his treatment of the subject states that "the directions perpendicular to the motion are obviously left unaffected by the transformation." In other words, he felt that the

$y' = y$ and $z' = z$ relations of the LT were indispensible, which is tantamount to making the same choice for ε as did Einstein, Lorentz, Poincaré and Larmor. Goldstein's assertion is no less compelling than the other justifications mentioned first, however. Other authors have insisted that eq. (11) is essential to guarantee adherence to the light-speed condition. This conclusion is clearly incorrect because of the homogeneous nature of this equation; multiplication of its right-hand side by any value of ε^2 satisfies the equation equally well. One should therefore remain open to critical examination of this point so that the question can finally be answered in a truly definitive manner.

The derivation of the LT has been carried out under the assumption that both rest frames are inertial. However, one should avoid the conclusion that it is only applicable when this condition has been met. This would make the theory all but useless since actual inertial systems are extremely rare. There are always forces acting on a given body, and achieving the perfect balance that inertiality demands is next to impossible and very difficult to maintain over any length of time. Physics is different than mathematics, however. One can postulate that the results of the LT can simply be extended to non-inertial systems. Then it is up to experiment to test whether this assumption is valid. Put another way, it is false to claim that the theory is not applicable to a given situation just because of the restrictive assumptions employed to derive it.

Chapter 3

Predictions and General Characteristics of the Lorentz Transformation

The Lorentz transformation (LT) is the basis of Einstein's Special Theory of Relativity (STR).[5] It leads directly to a number of experimental predictions about space and time relationships. The most prominent of these are FitzGerald-Lorentz length contraction (FLC) and time dilation. The LT is consistent with Einstein's first postulate of relativity, the relativity principle (RP), and it also guarantees that the speed of light is independent of the state of motion of both the observer and the light source, his second postulate. One of its most distinctive characteristics is the condition of Lorentz invariance, eq. (11) of the previous section. The latter's highly symmetric appearance clearly fulfills the requirements of the RP and its influence has gone well beyond the realm of classical relativity theory. The LT is also the basis for the space-time mixing concept that represents a revolutionary departure from Newton's fundamental principles. For example, it does away with the precept of absolute remote simultaneity of events. The relativistic velocity transformation (VT) is a corollary of the LT and is obtained from it by straightforward division of its space and time relations. However, it will be shown in the following discussion that it is important to distinguish between experimental predictions that require the LT specifically for their derivation, such as the FLC and time dilation, from those that flow directly from the VT alone, such as the aberration of starlight from the zenith and the results of the Fresnel light-drag investigation.

A. FitzGerald-Lorentz Length Contraction

The theoretical possibility of relativistic length contraction obtained credence in the latter part of the 19th century.[4] Experiments that were carried out in this period demonstrated that key assumptions of classical theory were in need of revision. FitzGerald[10] was the first to suggest that the negative result of the Michelson-Morley experiment[3] could be explained by assuming that the lengths of objects change as they move through the air. A few years later Lorentz[11] independently came to the same conclusion. Both authors believed that it was necessary to assume that there is an interaction with an "aether" that is responsible for the decrease in length of an object along the parallel direction. A key aspect of their speculation was that no change was expected in the perpendicular direction. One of the main objectives of Einstein's relativity theory in 1905[5] was to show that an aether is *not* required to explain the phenomena in question, but he nonetheless concluded on the basis of the LT that the FLC hypothesis was correct without any need for modification. It has proven quite difficult to verify this prediction of Einstein's theory on a definitive basis, but there have been some observations which have been claimed to at least provide convincing indirect confirmation of its validity.

The original deduction of the FLC is based squarely on the classical Galilean velocity transformation (GVT), which assumes that velocities are simply additive. The GVT by itself is unable to explain the experimental observation that the speed of light in free space is independent of the state of motion of the source. This was the result of the Michelson-Morley experiment,[3] for example, at least if one neglects the effects of the air through which the light moved.

Application of the GVT to the Michelson-Morley experiment is quite straightforward. It is assumed that light travels at speed c relative to the observer when the source is stationary in the laboratory. It is further assumed that there exists a medium/aether traveling at speed v along the positive x-axis. When the light is dragged along by the aether, the resultant speed of the light is dependent on its direction. It has a maximum value of c + v if the direction is parallel to that of the aether, but only c − v if the direction is anti-parallel. Due to Fizeau's experimental verification of the Fresnel light-drag effect in 1851, it was already known to physicists[12] that this view of the kinematics of light was oversimplified if the medium is water or some other transparent liquid. Nonetheless, it was concluded on the basis of the GVT that the time Δt_x required for a light beam to pass a distance L between the source and a mirror and then return to its starting point along the x-axis is:

$$\Delta t_x = L(c + v)^{-1} + L(c - v)^{-1} = 2Lc^{-1}(1 - v^2c^{-2})^{-1.} \tag{18}$$

It was further assumed that the speed of light was not affected by the aether if it moves in a perpendicular direction. Use of the Pythagorean Theorem then gives the following result for the return-trip elapsed time Δt along the y axis:[13]

$$\Delta t_y = 2Lc^{-1}(1 - v^2c^{-2})^{-0.5.} \tag{19}$$

The conclusion was directly applicable to the Michelson-Morley experimental arrangement.[3] If two light beams left the origin at the same time, traveling exactly the same distance L to their respective mirrors, they would not return at the same time if the speed of the aether was not equal to zero. The FLC assumption is intended to remove any such discrepancy in the elapsed times. Accordingly, the distance L in eq. (18) was predicted to be contracted by a factor of $\gamma^{-1} = (1 - v^2c^{-2})^{0.5}$ when the light moves along the x-axis because of an interaction with the aether, whereas the distance was unchanged in eq. (19) when the light moves along a perpendicular direction. This correction succeeds in making $\Delta t_x = \Delta t_y$, in *ad hoc* agreement with the experimental result obtained by Michelson and Morley in their interference experiment.[3] Because of the RP, it had to be assumed that an observer co-moving with the aether would not notice any contraction in his rest frame, but FitzGerald nonetheless concluded[4,10] "that the molecular forces are affected by the motion, and that the size of a body alters consequently."

The FLC is derived from the LT of eq. (17) in the following way. The length of a rod which is at rest in S′ is to be measured by the observer at rest in S. The rod is first aligned along the x direction and its length is determined to be $L_0 = x_2' - x_1'$ by the local observer in S′. The LT predicts that the positions of the two termini of the rod measured in S satisfy the following relations at time t: $x_2' = \gamma(x_2 - vt)$ and $x_1' = \gamma(x_1 - vt)$. It is crucial that the observer make the two measurements at the same time, something which is arguably impossible since the rod is moving relative to his position. Nonetheless, the above two equations can be combined to give:

$$L_0 = x_2' - x_1' = \gamma(x_2 - x_1) = \gamma L, \tag{20}$$

where L is the length of the rod measured by the observer who is at rest in S. Since $\gamma > 1$, it follows that $L < L_0$ and therefore that the rod is contracted from his vantage point. The analogous result for the case when the rod is oriented along the y axis is:

$$L_0 = y_2' - y_1' = y_2 - y_1 = L, \tag{21}$$

i.e., there is no difference between the two measurements if the rod is oriented in a direction perpendicular to the relative velocity of S and S′. Both results are seen to be in complete agreement with the conclusions of FitzGerald and Lorentz discussed first.

The fact that the length contraction is not the same in all directions has evoked much discussion over the succeeding years. Another aspect of the derivation has also received considerable attention. If the situation is reversed so that an identical rod is stationary in S, one can use the inverse transformation of the LT to obtain the analogous length prediction for the observer in S′. The latter equations can be obtained by algebraic manipulation of eq. (17). The same result can be obtained by simply interchanging the primed and unprimed indices and reversing the sign of v:

$$\begin{aligned} t &= \gamma(t' + vc^{-2}x') = \gamma\eta'^{-1}t' \\ x &= \gamma(x' + vt') \\ y &= y' \\ z &= z', \end{aligned} \tag{22}$$

with $\eta' = (1 + vu_x'c^{-2})^{-1}$ and $u_x' = x't'^{-1}$. The relations for the two termini at a particular time t′ are thus $x_2 = \gamma(x_2' + vt')$ and $x_1 = \gamma(x_1' + vt')$, so that

$$L_0 = x_2 - x_1 = \gamma(x_2' - x_1') = \gamma L'. \tag{23}$$

Consequently, the length L′ measured in S′ is seen to be smaller than L_0. In other words, both observers find that the rod is contracted when they observe it in a moving rest frame. This result corresponds to a *subjective* view of the measurement process that stands in stark contrast to the long-held position of complete objectivity espoused by Newton and other classical physicists (see Appendix E). Both observers conclude that the other's rod is shorter even though they agree on the length when they are in the same rest frame as the rod. This is an example of what will be referred to in the following as *Einstein's symmetry principle* (ESP). A similar result holds in STR for elapsed times and inertial masses, and indeed for any other conceivable physical property.

The latter consideration raises the question of whether the variation of an object's length with its speed relative to the observer is a real effect. Does the rod change its length or not? The main question in settling this issue is how to measure lengths when the object is not stationary in the observer's rest frame. The answer is to use atomic clocks and Einstein's light-speed postulate, as discussed below.

B. Relativistic Time Dilation

An analogous procedure can also be applied for elapsed times. In this case two events occur at times t_1 and t_2 on the basis of a stationary clock in S. The time difference measured in S′ is $t_2' - t_1' = T_0$. According to the LT, if the two times are measured at the same position in S′ ($x_2' = x_1'$), it follows that, since $x_2 - x_1 = v(t_2 - t_1)$,

$$T_0 = t_2' - t_1' = \gamma(t_2 - t_1) - \gamma v c^{-2}(x_2 - x_1) = \gamma(t_2 - t_1)(1 - v^2c^{-2}) = \gamma^{-1}T. \quad (24)$$

Evidently, the clock in S′ runs slower than its counterpart in S. One refers to this as "time dilation" in the (moving) rest frame S′ from the vantage point of a stationary observer in S. One can use the inverse LT in eq. (22) to predict the corresponding relationship when the same time difference (T_0) is measured locally in S ($x_2 = x_1$); note that $x_2' - x_1' = -v(t_2' - t_1')$:

$$T_0 = t_2 - t_1 = \gamma(t_2' - t_1') + \gamma v c^{-2}(x_2' - x_1')$$
$$= \gamma(t_2' - t_1')(1 - v^2c^{-2}) = \gamma^{-1}T'. \quad (25)$$

Time dilation is again measured in the moving rest frame. This is another example of the ESP discussed above. The clock in a moving rest frame always runs slower, just as distances in the moving rest frame are always measured to be shorter.

There are two basic differences about the time dilation prediction of the LT *vis-à-vis* that of the FLC. First of all, Einstein made another prediction in his original work[5] about changes in clock rates that in an important way runs counter to the first one. He gave the example of an electron moving in a closed path and returning to its starting point. He argued that a clock moving with the electron would have *lost* time upon its return. This conclusion violates the ESP because there is no disagreement between the moving and the stationary observer as to which clock runs slower during the flight of the electron. He then gave the more practical example of two clocks at rest at the Equator and one of the earth's Poles, respectively. In agreement with his latter conclusion, he stated that the equatorial clock would run slower. Although he did not further elaborate on the basis for the expected violation of the ESP in this case, von Laue[14] later argued that the reason for the distinction is that one of the clocks undergoes acceleration while the other does not. This fact is inconsistent with the underlying assumption of the RP and therefore there is a clear means of distinguishing between them because the one at the Equator is in constant acceleration relative to the one located at the Pole. This raises the question of whether a similar distinction might occur for distance measurements. A related question is whether the LT is valid in the case where acceleration is involved.

The other difference between the predictions of time dilation and the FLC is that the former is clearly more accessible to experimental testing. There are two reasons for this assessment. First, a clock has "memory" whereas a measuring rod does not. It is easy in principle to decide which of two clocks has lost time as a result of its motion by making a comparison of elapsed times between them when they are later brought back together in the same rest frame. Even if a rod has undergone a change in length during its journey, it seems inevitable that when it returns to its original rest frame it will be impossible to prove experimentally that this has occurred. Moreover, it is at least conceivable that the rates of clocks can be compared while they are still in relative motion to another.

As will be discussed in Chapter 5, Einstein[5] predicted on the basis of the LT that there is a transverse Doppler effect for light frequencies. Accordingly, the measured frequency ν_r for a moving light source should always be less in the laboratory than the corresponding value ν_e obtained by an observer co-moving with the light source:[15]

$$\nu_r = \nu_e(1 - v^2/c^2)^{0.5} = \gamma^{-1}\nu_e. \qquad (26)$$

The expected result is clearly consistent with the LT-version of time dilation because it satisfies the ESP. A "two-way" experiment in which two observers in relative motion each measure the frequency of light emitted from an identical light source in the other's rest frame should be capable in principle of settling this issue definitively.

There is an interesting twist to this proposal, however. Because of the light-speed postulate, one expects that the corresponding wavelength λ of light is *inversely* proportional to its frequency ν. The analogy to eq. (26) for wavelengths is thus:

$$\lambda_e = \lambda_r(1 - v^2/c^2)^{0.5} = \gamma^{-1}\lambda_r. \qquad (27)$$

This formula indicates that the wavelength λ_r received in the laboratory is *greater* than the wavelength λ_e measured at the light source. As a result, one can speak of *length expansion* accompanying time dilation in the rest frame of the moving light source rather than the length contraction predicted by the FLC. Moreover, the effect should be the same in all directions, again contrary to what one expects on the basis of the FLC and the LT.

C. The Relativistic Velocity Transformation (VT)

The relationship between velocity components can be obtained by simple division of the three spatial equations of the LT by that for time in eq. (17). The same procedure is employed to derive the classical velocity transformation (GVT) from the Galilean transformation. Einstein also chose another path in his original work,[5] however. He showed that the product of two four-dimensional space-time transformation matrices corresponding to relative speeds v_1 and v_2 leads to a new matrix of exactly the same form for the speed v_3 defined as:

$$v_3 = (v_1 + v_2)(1 + v_1 v_2/c^2)^{-1}. \qquad (28)$$

This is the new additivity principle to replace the simpler version of eq. (3) for the GVT. As mentioned in Chapter 2. C, this result proves that the transformation matrices form a group, a fact that

Einstein also used to justify his choice of the normalization factor in the GLT of eq. (15). Poincaré had used the same argument independent of Einstein at about the same time.[4] It needs to be stressed, however, that this property of the LT only holds if v_1 and v_2 are in the *same* direction.

In order to derive the relativistic velocity transformation (VT), it is necessary to start with a form of the LT which is based on intervals of time and distance. One needs two sets of space-time measurements to gauge the speed of an object. Because of the condition that both observers as well as the object of their measurements are in uniform translation, it is clear that differences of these quantities ($\Delta x = x_2 - x_1$ *etc.*) satisfy the same transformation relations as the original for a single measurement. The GLT of eq. (15) in Chapter 2. B has already been expressed in the required form. The velocity components u_i of the object are then obtained by division as $u_x = \Delta x/\Delta t$, *etc.* There is no need to take the additional step of computing the limit of each quantity as $\Delta t \rightarrow 0$ because the velocity components are constants under these circumstances.

It is advantageous to employ the GLT to obtain the VT for another reason, however. The result of the division is clearly independent of the normalization parameter ε in eq. (15), so one obtains exactly the same result for any value of this quantity as when Einstein's value[5] of $\varepsilon = 1$ is used. With the above definitions the VT has the following form:

$$\begin{aligned} u_x' &= (u_x - v)(1 - vu_x c^{-2})^{-1} = \eta(u_x - v) \\ u_y' &= u_y[\gamma(1 - vu_x c^{-2})]^{-1} = \eta\gamma^{-1}u_y \\ u_z' &= u_z[\gamma(1 - vu_x c^{-2})]^{-1} = \eta\gamma^{-1}u_z, \end{aligned} \tag{29}$$

with $\eta = (1 - vu_x c^{-2})^{-1}$. The corresponding inverse transformation is obtained by interchanging the primed and unprimed symbols and reversing the sign of v:

$$\begin{aligned} u_x &= (u_x' + v)(1 + vu_x' c^{-2})^{-1} = \eta'(u_x' + v) \\ u_y &= u_y'[\gamma(1 + vu_x' c^{-2})]^{-1} = \eta'\gamma^{-1}u_y \\ u_z &= u_z'[\gamma(1 + vu_x' c^{-2})]^{-1} = \eta'\gamma^{-1}u_z, \end{aligned} \tag{30}$$

with $\eta' = (1 + vu_x' c^{-2})^{-1}$.

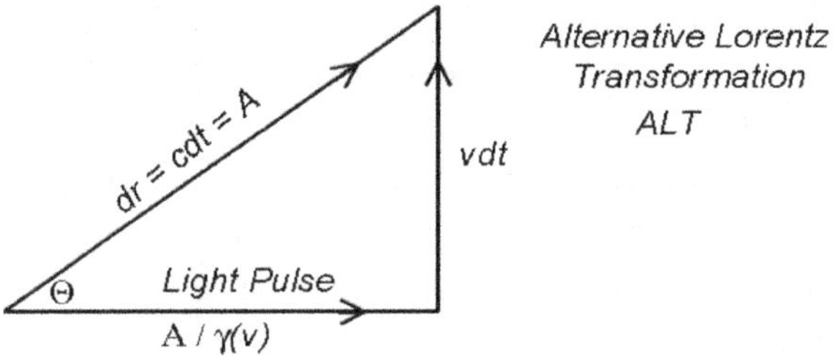

$$\gamma(v) = (1 - v^2/c^2)^{-1/2}$$

$$dt = A/c$$

$$dr^2 = A^2\gamma^{-2} + v^2 dt^2 = c^2 dt^2 = A^2$$

$$\tan\Theta = \gamma v dt / A = \gamma\, v/c$$

Fig. 2. Diagram showing a light pulse traveling in a transverse direction on a satellite that is moving with speed v relative to observer O. The distance dr = A traveled by the light pulse is computed from O's vantage point by employing the alternative Lorentz transformation (ALT) discussed in Chapter 6. A. He finds that the light pulse travels at an angle $\Theta = \tan^{-1}(\gamma\, v/c)$ and with speed equal to c, in agreement with the VT of eq. (29), so that his measured elapsed time is dt = A/c. Another observer (M) who is stationary on the satellite, who has the same unit of time as O (Q = 1 in the ALT), *i.e.* the same clock rate, also finds the light pulse to travel the same distance from his perspective (dr' = A = cdt') at the same speed c, but in a different direction (Θ = 0). The elapsed time for this event measured by M is therefore dt' = A/c, the same value as for O. The light pulse therefore arrives at the same time at the detector on the satellite for the two observers, despite the fact that they are in relative motion to each other. In the general case ($Q \neq 1$), the respective values for M are dt' = A/Qc and dr' = A/Q. Note that the light pulse still arrives at the detector at the same time for both observers after taking account of the difference in their respective clock rates.

Note that the first relation in eq. (30) is exactly the same as eq. (28) except for a difference in notation. It should be emphasized that the group theoretical argument Einstein and Poincaré used to derive eq. (30) is specific to the LT with $\varepsilon = 1$. Multiplication of two GLT matrices gives a product matrix containing ε^2 rather than the required value of ε needed to satisfy the definition of a group.

One of the most important applications of the VT is for the aberration of starlight at the zenith. A light pulse moving along the y axis toward the earth appears to be coming at an angle Θ from the normal (Fig. 2) because the star (S′) is moving with speed v relative to the laboratory (S) along the x-axis. Using the notation of eq. 30, $u_y{}' = c$ and $u_x{}' = 0$ for the light pulse from the vantage point of an observer at rest in S′. The value of η′ is therefore unity, so the VT predicts that the observer in the laboratory on earth will find that the components of the light pulse's speed are $u_x = v$ and $u_y = c\ (1 - v^2c^{-2})^{0.5} = c\gamma^{-1}$. The light appears to be coming at an angle $\Theta = \arctan(u_x/u_y) = \arctan(\gamma v/c)$, in agreement with experiment, with the required speed of $(u_x{}^2 + u_y{}^2)^{0.5} = c$. The value for Θ expected from the classical GVT is arctan (v/c), which is simply the low-v limit of the VT result.

It is interesting to also apply the inverse transformation of eq. (29) to obtain the velocity components of the light pulse in the star's rest frame from the above values measured on earth, *i.e.* $u_x = v$ and $u_y = c\gamma^{-1}$. The value of η is $(1 - vu_xc^{-2})^{-1} = \gamma^2$. Thus, one obtains $u_x{}' = 0$ and $u_y{}' = \eta\gamma^{-1}u_y = \eta\gamma^{-2}c = c$, as required. It will subsequently be shown (see Appendix B) that $\eta\eta' = \gamma^2$ in the general case, not only for light pulses.

The success of Einstein's theory in predicting the aberration of starlight at the zenith is generally looked upon as a confirmation of the LT. There is an important distinction to be made in this respect, however. The same result is obtained when the GLT of eq. (15) is used for any value of the normalization factor ε. From a strictly logical point of view, these results only serve as a direct confirmation of the VT. They therefore leave open the possibility that any other value of ε in the GLT is equally likely to correspond to the physically correct space-time transformation. This point will be taken up in more detail in Chapter 6.

One of the most influential theories of light in the early 19th century was put forward by Fresnel. It was based on the aether concept and one of its main predictions was that light traveling through a medium which itself was in motion in the laboratory would be subject to a drag effect. Although Einstein did not mention this experiment in his original paper,[5] von Laue showed a few years later[16] that its results could be explained quantitatively on the basis of the VT.

Experiments carried out by Arago in 1810 indicated that the resultant speed of the light could not be obtained in the standard way by simply adding the speeds of the medium and that of light in free space. Fizeau carried out more accurate studies of this effect in the 1850s and his results confirmed the light-drag effect, albeit of a smaller magnitude than had originally been thought.

On this basis it was widely concluded that the only way to explain the slowing down of the light was by accepting as fact that it moves through a stationary aether. The dependence of the speed of light c′ on the speed v of a transparent medium characterized by refractive index n through which it is transmitted was determined by Fizeau to be:[4]

$$c' = cn^{-1} + v(1 - n^{-2}), \qquad (31)$$

where $c = 2.99792458 \times 10^{8}\ ms^{-1}$ is the speed of light in free space (an additional wavelength-dependent term equal to $-v\lambda n^{-1}dn/d\lambda$ also results from consideration of the Doppler effect and has been observed[17]). According to the classical Galilean transformation, c′ should simply be equal to $cn^{-1} + v$, *i.e.*, the sum of the speed of light in the refractive medium and that of the medium itself. The $(1 - n^{-2})$ factor of v in eq. (31) is less than unity for normally refractive media (with $n > 1$) such as water, hence the coining of the term "light drag" to describe the effect.

Von Laue[16] was able to successfully derive Fizeau's experimental result simply by using eq. (30). One obtains the desired result in eq. (31) by setting $u_x' = cn^{-1}$ in the VT and neglecting higher-order terms in the expansion of η′:

$$u_x = c' = (1 + vn^{-1}c^{-1})^{-1}(cn^{-1} + v) \approx cn^{-1} + v - vn^{-2}. \qquad (32)$$

The true speed of light in the medium at rest in the laboratory is not inversely proportional to n but rather to the group index of refraction n_g, whereby the latter is defined by the expression:[18,19]

$$n_g = n + \omega dn/d\omega, \qquad (33)$$

where ω is the circular frequency of light. The refractive index n varies with ω due to what is conventionally described as a dispersion effect. As a result, the experimental value[20] for the light speed in the medium at rest is given as

$$c'(v = 0) = cn_g^{-1} = cn^{-1} - kcn^{-2}dn/dk, \qquad (34)$$

where $k = \omega nc^{-1}$.

There is another piece of information to be gleaned from eq. (31) that seems to have eluded the attention of 19th century physicists, however, and continues to be largely ignored to the present day. If one examines the limiting situation for light traveling in free space (n = 1), it is seen that the speed dependence of the medium disappears entirely, with the simple result:

$$c'(v) = c. \qquad (35)$$

This is an experimental verification, or at least a straightforward and unavoidable deduction based on experiment, of Einstein's light-speed postulate.[5] It shows that Fizeau's experiment could just as well have been taken as motivation to confirm eq. (34) in different empirical investigations as it was to prove the existence of an aether. As it turned out, as for example with the Michelson-Morley experiment,[3] attempts to do the latter kept coming back to the same, thoroughly surprising to the great majority of contemporaries, result that the speed of light actually does not depend on the relative speed of the source to the observer, exactly as Fizeau demonstrated with his light-drag experiment.

Contemporary experiments[21] with the Sagnac effect[22] have tended to cloud the issue, however. The point of contention is perhaps best illustrated by the operation of the Global Positioning System (GPS) in computing distances on the earth's surface. For this purpose the speed of light must be adjusted to take account of the receiver speed v,[23] *e.g.* it is taken to be c + v if the receiver is moving towards the satellite and c – v if it moves in the opposite direction. The object is to measure the distance D between the satellite and the receiver *at the time that the light signal is emitted.* If the receiver is moving away from the satellite with speed v, however, the signal must travel farther than D in order to be received. If the time to reach the receiver is Δt, this means the actual distance travelled by the light is $D + v\Delta t$, not simply D. According to Einstein's second postulate,[5] the signal traverses this distance with speed c, so that the elapsed time Δt is equal to $(D + v\Delta t)c^{-1}$. The value of the required distance between the satellite and the *original* position of the receiver is thus $D = (c - v)\,\Delta t$. The "adjustment" of the speed of light in the GPS procedure is thus seen to be a consequence of the fact that the light must travel a different distance than D to reach the receiver, *and not because the speed of light is dependent on the speed of the receiver*. The GPS effect is actually a confirmation of the second postulate and a further demonstration that the corresponding Galilean-transformation result is contradicted by experiment. A similar argument applies to the original Sagnac effect[22] observed for the interference of two light waves moving in opposite directions on a rotor. There is a time discrepancy Δt caused by the fact that the waves must travel different distances before arriving at the same point on the rotor. The observed value of Δt is obtained by assuming that the light speed is c for both waves.[24]

Perhaps the main function of the VT is to guarantee adherence to the light-speed postulate. This property is obvious from the way the GLT and VT have been derived above. It is instructive to just choose random values for u_x, u_y and u_z in eq. (29) which correspond to a magnitude/speed for **u** that is equal to c (note that the Pythagorean Theorem can be used to determine the magnitudes of vectors in all such applications). The result for **u′** always has the same magnitude (c) even though the direction of this vector is not generally the same as for **u**. An example of this type has already been considered above in the discussion of the aberration of starlight at the zenith.

It is also instructive to consider the VT equations when they are expressed in polar coordinates:[25] $u_x = u \cos \vartheta$, $u_y = u \sin \vartheta \cos \phi$, $u_z = u \sin \vartheta \sin \phi$ and corresponding relations for u_x', u_y', u_z' in terms of ϑ' and ϕ', respectively. The speed u depends on the input specifications for the observer in S′ as follows:

$$u = (u_x^{\ 2} + u_y^{\ 2} + u_z^{\ 2})^{0.5} = [u'^2 + v^2 + 2vu'\cos \vartheta' - (vu'c^{-2}\sin\vartheta')^2]^{0.5}. \quad (36)$$

The corresponding relationship between ϑ and ϑ' is:

$$\vartheta = \arctan\, (u_y^{\ 2} + u_z^{\ 2})^{0.5} u_x^{\ -1} = \arctan\, [\gamma^{-1}(v + u' \cos \vartheta')^{-1} u'\sin\vartheta']. \quad (37)$$

The latter equation reduces to the formula for the angle of aberration of starlight at the zenith given above (with $u' = c$ and $\vartheta' = \pi/2$): $\Theta = \arctan\, (u_x/u_y) = \arctan\, (\gamma v/c)$. The two azimuthal angles ϕ and ϕ' are equal according to the VT, because $u_z/u_y = u_z'/u_y'$. The transformation does not produce any twisting when the polar axis is chosen to be parallel to the x-axis in eqs. (29-30).

Rearrangement of eq. (36) leads to another useful property of the VT that relates $\gamma(u)$ to $\gamma(u')$ and $\gamma(v)$:

$$(1 - u^2c^{-2})^{0.5} = (1 - v^2c^{-2})^{0.5}(1 - u'^2c^{-2})^{0.5}(1 + vu'c^{-2}\cos\vartheta')^{-1}. \quad (38)$$

If $u' < c$, the denominator on the right-hand side is always positive as long as the relative speed v of S and S′ is also less than c. It can therefore be concluded that $u < c$ as well under these conditions, *i.e.*, because $\gamma^{-1}(u) > 0$. It is also obvious that $u = c$ whenever u′ = c and v < c. One also sees that if v = u′ and ϑ′ = π, then u = 0, *i.e.* the object is stationary in S. On the other hand, if u′ > c and v < c, both the left- and right-hand sides of eq. (38) have imaginary values, which means that u > c. Evidently u = u′ = c is a definite barrier that cannot be transcended by simply transforming to a different inertial frame as long as v < c.

The VT has also proven quite instructive in particle physics.[25] For example, consider a slightly altered version of eq. (37) below:

$$\tan \vartheta = \gamma^{-1} \tan \vartheta'(1 + v\, u'^{-1} \sec \vartheta'). \qquad (39)$$

There is a range of v > u′ such that particles moving backwards in S′ (ϑ′ = π) move forward in S (tan ϑ/tan ϑ′ < 0).

D. Thomas Spin Precession

The VT can also be used to describe processes that involve acceleration by applying it at successive time intervals where the velocities of the object are **v** and **v** + **a**dt. The orbiting electron in an atom is a good example, in which case the system undergoes a sidewise acceleration **a**. In this case the LT has to be applied with two velocities that are not parallel to one another[4] and thus do not conform to eq. (28). The spin direction is constantly changing even though there is no external torque.[26] Thomas showed[27] that the change dϕ in the angle between the velocity **v** of the electron and that of the electron spin is given by

$$\begin{aligned} d\varphi &= -(\gamma - 1)\, d\vartheta = -(\gamma - 1)\left|\mathbf{v} \times \mathbf{a}\right| dt\, v^{-2} \\ &= -\gamma^2 \left|\mathbf{v} \times \mathbf{a}\right| dt \left[(\gamma + 1)c^2\right]^{-1} \end{aligned} \qquad (40)$$

While the velocity of the electron turns in the orbital plane by dϑ, the spin projection turns in the opposite sense.[26] The angular velocity ω_T of the Thomas precession is thus:

$$\omega_{\mathrm{T}} = d\varphi / dt = \frac{-\gamma^2 \left|\mathbf{v} \times \mathbf{a}\right|}{(\gamma + 1)c^2} \qquad (41)$$

It can be noted that eq. (41) results from use of the GLT of eq. (15), and not only the LT. This is because both dφ and dt are proportional to the normalization factor ε in the GLT,[28] which is therefore cancelled in obtaining ω_T.

Eq. (41) can also be applied to a body spinning in orbit around the earth,[26] as will be discussed subsequently when attention turns to the study of gravitation. In the Newtonian theory,

$\mathbf{a} = -GM\mathbf{r}/r^3$, where G is the universal gravitational constant, M is the gravitational mass of the earth and **r** is its radius vector to the earth's center of mass. The theory of general relativity predicts[29] an effect which is three times larger and of the opposite sense. Unfortunately, no experimental test has yet been carried out to decide which theory is correct.

Chapter 4

Checks for Internal Consistency of the Lorentz Transformation

The standard way to judge the validity of a physical theory is by comparing its theoretical predictions with experimental data. However, especially when the subject of the theory is as complex as relativity, there is another characteristic that also needs to be carefully analyzed. Any physical theory is based on a number of assumptions that cannot be justified on the basis of what are commonly called "first principles." Newton's Laws of Kinematics provide a good example. There is no *a priori* justification for assuming that a body will continue in constant motion as long as it does not experience an outside force. There is also no proof for the principle of energy conservation. When there is more than one such assumption, however, there is always the theoretical possibility that they are somehow inconsistent with one another.

It is the nature of physical theories that it may not be obvious where to look for such deficiencies, so one can say that such a search can never really be brought to a conclusive end. Showing that the theory withstands such scrutiny for a wide variety of applications is encouraging, but nothing more than this from a purely logical point of view. *The opposite is true, however.* Once an inconsistency is found that is the necessary consequence of the premises of the theory, there is no recourse but to alter it in such a way as to remove the contradiction. The challenge is to find the offending assumption and replace it with something else that accomplishes this goal without yielding on any of the proven successes of the previous version of the theory.

One of the advantages of the Special Theory of Relativity (STR) is that it is based on two clearly stated assumptions, the RP and the light-speed postulate, both of which are firmly grounded upon a wealth of solid experimental findings. It therefore seems highly unlikely that STR could be self-contradictory for the very reason that its premises are so well-founded. It is nonetheless worthwhile to make the effort, as will be shown below.

A. The Symmetry Principle of STR

There have been a great many attempts to disprove Einstein's theory of relativity on the basis of its prediction that two clocks in motion can both be running slower than each other at the same time.[30,31] This characteristic has been referred to in Chapter 3. A as Einstein's Symmetry Principle (ESP). It applies equally well to other quantities such as lengths and inertial masses. The general argument is based on a well-known fact about measurement, namely that the magnitudes of physical properties of the same kind can always be ordered. To say that clock A runs slower than clock B implies that clock B is faster than clock A. At least, that is the situation in normal experience when the clocks are sitting side-by-side in a laboratory.

The situation with the ESP for clocks is different, however. By construction, the clocks are moving relative to one another. It is not possible under these circumstances for a single observer to make a direct comparison of their rates in the usual sense of the word. It might sound absurd that this distinction could be decisive, but one cannot rule out the possibility based on purely logical principles. One thing becomes clear, however. If the ESP is correct, it means that the measurement process is not totally objective. The ratio of the clock rates is a matter of perspective in this situation. It violates an ancient law of physics which will be referred to in the following as the Principle of Rational Measurement (PRM). Two observers may disagree as to the numerical values of two quantities of the same type by reason of the fact that they express their measured results in a different set of rational units, but *they must agree on the ratio of their respective numerical values*. Indeed, it becomes completely impractical to define a suitably universal set of units that both can use in a definitive manner if the two observers cannot even agree on which of the two properties is larger or smaller.

From the above discussion it might be concluded that even the possibility of a definitive experimental test is excluded because of the need to measure the rates of the clocks while they are still in relative motion to one another. The transverse Doppler effect discussed in Chapter 3. B offers a practical means of overcoming this obstacle, however. It is generally agreed that the observation of a red shift [see eq. (26)] in the frequency emitted from a moving light source is indisputable proof that the clocks in that rest frame run uniformly slower than in the laboratory where the actual measurement is carried out. Theoretically, one can test the ESP by carrying out a two-way experiment in which light signals are exchanged between two observers in different inertial systems. If a red shift is indeed found in both directions, this could certainly be taken as strong evidence, if not a strict proof, that the ESP does occur in nature.

There is another way of approaching this question, however. Consider the following application of the ESP to time dilation[30,31]. A collection of meta-stable particles has a lifetime of 1.0 μs in the laboratory S. When identical particles are accelerated to another inertial system S′, the laboratory observer in S finds because of time dilation in the latter rest frame that their lifetime has increased to 10.0 μs. Behavior of this type has been observed, as will be discussed in Chapter 5.B. According to the ESP,[5] however, an observer at rest in S′ must find that the particles stationary in S have a lifetime which is ten times longer than those in his rest frame. This means that according to the clocks in S, the particles must have a lifetime of 100.0 μs. Yet, these are exactly the same particles that were found to have a lifetime of only 1.0 μs at the start of the example. In reality, nothing has changed for these particles, despite the fact that they are ascribed two different lifetime values according to the ESP. Clearly, the predicted lifetime must continue to increase by a factor of 100 every time this back-and-forth operation occurs for the two clocks. This is obviously an impossible situation to realize in actual experiments. As such, this example represents an unequivocal contradiction of the ESP. Moreover, it also proves that the LT itself is not internally consistent.

B. Deductions of the Lengths of Objects: The Clock Riddle

The measurement of distances presents another opportunity for testing the internal consistency of STR. On the one hand, one has the FitzGerald-Lorentz length contraction (FLC) prediction of the theory that comes originally from consideration of the results of the Michelson-Morley experiment,[3] as discussed in Chapter 3. A. However, because of the light-speed postulate there is also the possibility of measuring distances with atomic clocks. Consider the following example. The length of a metal bar is found to have a value of L m by observers O and O′ while they are together on the earth's surface. Then O′ boards a satellite with the bar and proceeds to fly off into space. At some point he reaches a constant velocity **v** relative to O which he maintains thereafter. O′ will continue to measure the length of the bar to still be L m, independent of its orientation to **v**, as required by the RP. The FLC states that O will obtain the same value of L m if the bar is oriented perpendicular to **v** [see eq. (21)], but will find it to be contracted by a factor of $\gamma(v)$ if it is aligned in the parallel direction, *i.e.* he will measure the length of the bar to be γ^{-1} L m in this case.

The time-dilation effect of STR stipulates that the proper clocks of O run faster than those of O′ by the same factor of γ. Despite this, both observers should continue to measure the speed of light to have a value of c ms^{-1}. This means that O′ will measure the elapsed time $\Delta T'$ for a light pulse to pass between the termini of the bar to have a value of Lc^{-1} s, but that O will find a larger value of $\Delta T = \gamma L c^{-1}$ s. The latter two timing results are clearly the same in all directions because clock rates are not affected by the orientation of the bar. That means that O must also find the length of the bar to be independent of its orientation to v, namely to have a value of $c\,\Delta T = \gamma L$ m; in short, the bar *expands* instead of contracting or staying the same length according to this method. This is a clear contradiction of the FLC. *It shows that STR is self-contradictory,* just as has been concluded above with the muon-decay tests of the ESP.

The above argument has been referred to in previous work as the "clock riddle"[32] to distinguish it from the far better-known "clock paradox" that deals with the asymmetry of time dilation. The clock riddle of itself does not disprove either the FLC or the ESP, but it does show that the LT is invalid. Both ways of measuring distance are perfectly consistent with the LT and so there must be something about it that cannot stand. *It indicates that an error has been made in deriving the LT.* Before considering what this might be, it is best to consider another characteristic of the LT that has heretofore been considered to be inviolate: space-time mixing.

C. Dependence of Clock-Rate Ratios

Before the advent of STR it was taken for granted that space and time are two fundamentally different quantities. This attitude gradually began to change as the discussion turned to the development of a new space-time transformation that was consistent with the revolutionary experimental results that began to emerge in the latter half of the 19th century. Special attention was devoted to the possibility that the correct value for the free parameter ε in the GLT of eq. (15) could simply be unity. The equation connecting the two measured elapsed times of two observers in motion becomes:

$$\Delta t' = \gamma(\Delta t - vc^{-2}\Delta x) = \gamma\eta^{-1}\Delta t. \qquad (42)$$

It is interesting to examine the above argument for the special case in which the two events simply refer to the initial and final coordinates of an object in flight. In that case the ratio $\Delta x/\Delta t$ in eq. (42) is equal to the velocity component u_x of the object in the direction parallel to **v**, the relative velocity of S and S′. The corresponding ratio of elapsed times is thus:

$$\Delta t'/\Delta t = \gamma\eta^{-1} = (1 - v^2c^{-2})^{-0.5}\,(1 - vc^{-2}u_x). \qquad (43)$$

This ratio is determined solely by the relative rates of the proper clocks in S and S′, respectively. There is no reason why this ratio should depend in any way *on the speed of the object being measured by the two observers.* Hence, eq. (43) represents a violation of a basic physical principle, *i.e.* that extraneous factors should not influence the results of measurements. This result constitutes another proof that the LT is not a valid set of equations. It comes in addition to the problem connected with the clock riddle discussed in Chapter 4. B.

D. Remote Non-simultaneity

One of the most famous deductions that Einstein made based on his relativity theory is that two observers could disagree as to whether two events occur simultaneously or not. He rejected thereby the Newtonian concept of absolute remote simultaneity of events. The justification for Einstein's conclusion is found in the LT. Examination of eq. (15) shows that if $\varepsilon = 1$, as Einstein assumed in deriving the LT,[5] Δt and $\Delta t'$ can never be equal unless either Δx or v is equal to zero. In other words, if the two inertial systems S and S′ are in relative motion ($v \neq 0$) and the unprimed observer does not carry out his timing measurements for the two events at the same point in space ($\Delta x \neq 0$), then according to the LT, it follows that $\Delta t \neq \Delta t'$. This in turn means that *under these conditions*, it is impossible for the two events to be simultaneous in S and also be simultaneous in S′ (or *vice versa*). Poincaré[8] had expressed similar reservations about absolute simultaneity in 1898.

One of the most famous attempts to illustrate non-simultaneity involves a speeding train moving with speed v as it passes through a station. Two lightning strikes are observed to occur simultaneously at separate positions A and B as the midpoint M of the train passes an observer S on the station platform. The distance between A and M on the one hand and B and M on the other is the same (L m) from the standpoint of S. A light pulse from the lightning strikes reaches M in Lc^{-1} s from both A and B, so the events are clearly simultaneous for S (assuming that the speed of light is equal to c in both directions for him). A calculation[31] is then presented to show that it takes less time from the vantage point of a rider R on the train for the light pulse moving in the forward direction from A to arrive at M than it does for the other light pulse moving from B in the opposite direction. Over the past century this argument has received wide acceptance as proof that events can indeed be simultaneous for one observer but non-simultaneous for another who is moving with respect to him.

An interesting fact about the above derivation of non-simultaneity is that it does not make use of the VT. Since the train and the light pulses move in the same direction, only the first relation in eq. (30) is needed for this purpose. The observer R on the train finds that the two light pulses move in opposite directions with speed c, in agreement with Einstein's second postulate. Since the distance traveled is equal in each case, it follows that the two light pulses arrive at M simultaneously for R. To find the corresponding speeds for observer S, it is necessary to set $u_x' = c$ for the light pulse moving in the forward direction from A to M and

$u_x' = -c$ for that moving in the opposite direction from B to M. The speed v of the train is also needed. Substitution in the above equation gives for the forward-moving light pulse: $u_x = (1 + vc^{-1})^{-1} (c + v) = c$. The corresponding substitution for the backward-moving light pulse from B to M gives: $u_x = (1 - vc^{-1})^{-1}(-c + v) = -c$, *i.e.*, the same speeds as measured by R in both cases. Because of time dilation on the train, it follows that the elapsed time $\Delta t'$ measured for each light pulse to arrive at M is $\gamma(v)$ times slower than the time Δt measured for the same two events to occur from the standpoint of S. Nonetheless, the resulting two values of $\Delta t = \gamma(v)\ \Delta t'$ are equal, proving that the two events also occur simultaneously for S (a similar example for light pulses on an airplane is discussed elsewhere[33]). In short, use of the VT and strictly proportional time dilation, as assumed in Einstein's asymmetric theory,[5] leads to the opposite conclusion as that made in the first argument above, namely that the two events are indeed simultaneous for both R and S.

There is no reason why the VT cannot be used to solve this problem. It has had numerous successful applications, as discussed in Chapter 3.C. Moreover, it is completely unrelated to the ESP, which is a product of the LT alone, and therefore avoids the contradictory result inherent in the ESP. The fact is that the derivation using the VT is simply another example of a lack of consistency in the version of relativity theory developed by Einstein in his 1905 paper.[5] There should be no argument that a theory that gives opposite answers for the same problem depending on how it is applied cannot stand in its unmodified form. The problem can be solved by simply accepting the validity of the VT and the asymmetric theory of measurement and eliminating the LT and the ESP as a direct consequence.

It might be thought that the above proof of simultaneity is specific for light pulses, but it is easy to show that it has universal validity for all objects/particles. Assume that the observer R on the train measures the following results for two particles of *non-zero* rest mass moving in opposite directions, similarly as before with the light pulses: the speed of the midpoint of the train is $v_M = 0$, while the speed of the particle moving from A to M is $w < c$. Next calculate the *relative* velocity v_{AM} of the particle to the midpoint M. Note that it has the same value whether one uses the Galilean velocity transformation (GVT) or the relativistic velocity transformation (VT), namely v_{AM} (R) = w. Next calculate the speed of the particle from the viewpoint of observer S on the platform. He sees the midpoint M moving at speed v. The speed of the particle using the GVT for S is $v_A(S, GVT) = v + w$.

The speed of the particle using the VT for S is different, namely $v_A(S, VT) = (v + w)(1 + vwc^{-2})^{-1}$. This is only an intermediate value, however. What is needed is the speed v_{AM} (S) of the particle moving *relative* to the midpoint M [which travels with speed v for S, *i.e.* $v_M(S,VT) = v$]. For this purpose both the GVT and the VT will be used. The GVT result is easy, namely $v_{AM}(S, GVT) = (v + w) - v = w$. Note that this is the same result as found above for R: $v_{AM}(R, GVT) = v_{AM}(R, VT) = w$. The last step is to calculate v_{AM} (S,VT), that is, the difference $v_A(S, VT) - v_M(S, VT)$. The VT must be used to do this, however. The numerator is $N = [(v + w)(1 + vwc^{-2})^{-1} - v]$. There is a denominator D as well, namely, $D = [1 - (w + v)\ (1 + vwc^{-2})^{-1}vc^{-2}]$. Dividing N by D gives after cancellation of $(1 + vwc^{-2})$:

$$\begin{aligned} v_{AM}(S,VT) &= N/D \\ &= [v + w - v(1 + vwc^{-2})]c^2[(1 + vwc^{-2})c^2 - vw - v^2]^{-1} \\ &= (c^2v + c^2w - c^2v - v^2w)[c^2 + vw - vw - v^2]^{-1} \qquad (44) \\ &= w(c^2 - v^2)(c^2 - v^2)^{-1} = w. \end{aligned}$$

The end result is very simple and is also quite useful in refuting Einstein's claim of remote non-simultaneity:

$$v_{AB}(S,GVT) = v_{AB}(S,VT) = v_{AB}(R,GVT) = v_{AB}(R,VT) = w. \qquad (45)$$

The result is the same whether we use the GVT or the VT to obtain it. In short, the GVT and VT are in perfect agreement as to the speed of the object relative to the midpoint of the train, whether it be a light pulse or a bullet or something else entirely.

The application to simultaneity is obvious. If two observers agree on the speeds of the two particles at opposite ends of the train, they must also agree that they arrived simultaneously at its midpoint M. They will only disagree on the elapsed time for this to occur. If R finds that $T_A = T_B$ on his clock for both events, S will find that the corresponding times on his clock are γT_A and γT_B, respectively. Since the latter two elapsed times are equal, the two events are also simultaneous for S.

E. Einstein's Dual Theories of Relativity

In summary, STR in general, and the LT in particular, fail to demonstrate the required degree of internal consistency for a physically valid theory, despite the many years of firm belief in Einstein's theory. Analysis of the above discussion shows that he laid the basis for dual versions of relativity theory in his original work. One version is based on the LT and is inextricably linked to the ESP and consequently on a *subjective* view of the measurement process. The other relies exclusively on his light-speed postulate and the asymmetric theory of time dilation he also mentions in his original work.[5] It is easily distinguished from the former version because it conforms to a strictly objective theory of measurement. Both versions are consistent with the VT and the many successful STR applications that flow directly from applying it.

The problem with having two distinct versions of the theory comes when attention is turned to length variations. The standard version based on the LT leads directly to the FLC and its prediction that length contraction by varying amounts accompanies the slowing down of clocks in a moving rest frame. If only the light-speed postulate is employed, however, the exact opposite conclusion results, namely that lengths expand isotropically in the moving rest frame in the same proportion as time dilation occurs for local clocks there. The same disconnect is found in the train example, in which case the standard theory predicts non-simultaneity of events for the two observers, whereas exclusive use of the light-speed postulate finds that everything occurs simultaneously for them. The explanation used to explain the fact that the two versions of the theory obtain different answers for the symmetry, or lack thereof, in measurements of time dilation, namely that acceleration of one of the clocks is not consistent with the underlying assumptions of the LT, cannot be applied in these cases. The results of isotropic length expansion and remote simultaneity that follow from application of the light-speed postulate occur just as well when all objects are in states of pure translation. As a result, it is clear that only one of the dual theories can be valid. The contradiction inherent in the ESP demonstrates unequivocally that it is the LT version that must be discarded. In the following section it will be shown how these purely theoretical objections to STR are mirrored in a number of experimental investigations that have been carried out over the past century.

Chapter 5

Experimental Tests of Special Relativity Theory

The first experiment that provided a direct test for the existence of time dilation was carried out by Ives and Stilwell in 1938.[34] The transverse Doppler effect was measured for a moving beam of hydrogen atoms whose radiation was observed at small angles in both the incoming and outgoing directions.[34,35,36] This work was followed shortly thereafter by lifetime measurements of accelerated meta-stable particles (mesons). Both experiments verified the existence of time dilation.

A. Measurements of the Transverse Doppler Effect

The rationale for this approach was based directly on the STR formula for the relativistic Doppler effect in general. Accordingly, when light of wavelength λ_0 is emitted from a source moving with speed v relative to the laboratory, the observed wavelength λ at an angle Θ relative to the beam direction is given by the formula:[36]

$$\lambda = \gamma(v)\lambda_0(1 - vc^{-1}\cos\Theta) \approx \lambda_0(1 - vc^{-1}\cos\Theta + v^2c^{-2}/2). \qquad (46)$$

By measuring λ for light emitted in diametrically opposite directions and averaging the two results, the Θ-dependence can be eliminated and therefore the goal of determining the wavelength for transverse radiation ($\Theta = \pi/2$) can be achieved to a good degree of approximation, namely as $\lambda_Q \approx \lambda_0(1 + v^2c^{-2}/2)$.

In the most accurate experiment of this type, Mandelberg and Witten[36] reported a shift to longer wavelength of 0.219 Å for λ_0 = 6562.793 Å and a relative speed of v = 0.008176c, in excellent agreement in both magnitude and direction with the theoretical result above. It was also perfectly consistent with Einstein's prediction of time dilation since it indicated that the period τ of the moving clock/light source also increased by the same factor $\gamma(v)$ when it was accelerated relative to the laboratory:

$$\tau = c^{-1}\lambda_Q = \gamma c^{-1}\lambda_0 = \gamma\tau_0. \quad (47)$$

There are several aspects of these experiments that need more careful consideration. First, they do not allow for a definitive test of the ESP for time dilation because they are only carried out in one direction. Second, the results are also quantitatively consistent with Einstein's asymmetric theory of time dilation,[5] since it is the rate of the accelerated clock that is shown to decrease by virtue of its motion in the laboratory. However, the most significant detail of the transverse Doppler experiments is that it is the wavelength of the light that is measured directly in them and therefore that they also tell us something about the way the lengths of objects vary with their state of motion. Even more interesting is that the results show that the wavelength increases on the accelerated light source. At first glance, this might sound like a trivial point, simply verifying that the speed of light in free space, *i.e.* the product of wavelength and frequency, is the same in all inertial systems. This conclusion overlooks a straightforward deduction that can be made on the basis of the RP, however. A stationary observer in the rest frame of the light source cannot notice any change from the standard wavelength. This can only mean that all measuring devices in that rest frame have also increased in dimensions by exactly the same γ factor as for the wavelength. Moreover, the length expansion is the same in all directions (isotropic).

The increase in wavelength is consistent with what is expected in Chapter 4. B when distances are measured with clocks. The transverse Doppler experiments complement the arguments of the clock riddle.[32] They show unequivocally that the FLC is responsible for the contradiction. They still leave open the possibility that the light-speed postulate is correct and can be used to infer the dimensions of objects that are moving at high speeds (in S′) relative to the observer (in S). The length of the rod has expanded, independent of its orientation to **v**, by the same factor $\gamma(v)$ as the rates of co-moving clocks have decreased:

$$\Delta t' = \gamma^{-1}\Delta t,$$
$$\Delta x' = \gamma^{-1}\Delta x, \qquad (48)$$
$$\Delta y' = \gamma^{-1}\Delta y,$$
$$\Delta z' = \gamma^{-1}\Delta z.$$

This is in opposition to what the FLC predicts (see Chapter 3.A) when clock rates slow, namely: $\Delta t' = \gamma^{-1}\Delta t$, $\Delta x' = \gamma\Delta x$, $\Delta y' = \Delta y$ and $\Delta z' = \Delta z$.

Since the clock riddle proves that the LT is not valid, there is no compelling reason to believe in any of its predictions. Nonetheless, this state of affairs naturally does not rule out the possibility that one or more of them might still be genuine. The ESP is a prime example, and so there is certainly merit in carrying out further experiments which, unlike the Ives-Stilwell investigation,[34] are capable of giving a definite answer. A two-way experiment is required for this purpose, and this goal was achieved in 1960 by Hay *et al.*[37] These authors mounted an x-ray source and an absorber on an ultracentrifuge and measured the transverse Doppler frequency shift using the Mössbauer effect. The experimental arrangement had the advantage of guaranteeing that the radiation was emitted in the transverse direction at all times, thereby removing any uncertainty in the value of Θ in eq. (46). The absorber was located farther from the rim of the rotor than the light source, so the "observer" moves faster in the laboratory in this experiment. If the ESP is correct, this difference relative to the original Ives-Stilwell investigation should have no effect on the results and a *red shift* should be found in accordance with eq. (26). If Einstein's asymmetric theory of time-dilation holds instead, a *blue shift of the same magnitude* should be recorded.

Angular velocities ω of up to 500 revolutions per second were employed in the study of Hay *et al.*[37] Their results for the fractional shift in the light frequency are found to be in quantitative agreement with the formula: $0.5\ (R_1^{\ 2} - R_2^{\ 2})\omega^2 c^{-2} = 2.44 \times 10^{20}\omega^2$. They refer to this as the "expected shift" and claim that it can be derived in either of two ways from theory: a) from Einstein's equivalence principle[38] (EP) by treating the acceleration of the rotor as an "effective gravitational field" or b) from the time- dilation effect of STR. They do not comment directly as to whether the observed direction of the shift is to higher or lower frequency, but since the factor in the above equation is positive, a shift to the blue is clearly indicated if $R_1 > R_2$. As a consequence, this result stands in contradiction to the prediction of STR that the sign of the Doppler shift should always be the same as found in the Ives-Stillwell experiment,[34,35,36] namely in the direction of longer wavelengths and lower frequencies. Similar experiments with rotors were subsequently carried out by Kündig[39] and Champeney *et al.*,[40] with the same results. Kündig summarized his findings succinctly by stating "that the clock which experiences acceleration is retarded compared to the clock at rest." This means that the *absorber clock at the rotor's rim has the slower rate* and that a blue shift must therefore have been observed.

The symmetry characteristics of the Hay *et al.* empirical formula are particularly telling in this respect. The parameters R_1 and R_2 clearly refer to the distances of the absorber and source from the axis of the rotor, although no explicit designation is given in the text.[37] Since the former distance (R_a) is greater than the latter (R_s), it is clear from the sign on the right-hand side of the Hay *et al.* empirical formula that $R_1 = R_a$ and $R_2 = R_s$. The dependence of the fractional Doppler shift $\Delta\nu/\nu$ on ω is thus:

$$\Delta\nu/\nu = 0.5(R_a^{\,2} - R_s^{\,2})\omega^2c^{-2}. \qquad (49)$$

The above formula therefore implies that interchanging the positions of the absorber and light source on the rotor's axis causes a reversal in the sign of the Doppler shift. This is clearly not the result that one expects from STR and eq. (26), since it claims that only the relative speed $|R_a - R_s|\,\omega$ of the source and absorber is material in making this determination. According to the time dilation formula of eq. (47), the empirical results for transverse motion should obey the formula:

$$\Delta\nu/\nu = \gamma^{-1}(|R_a - R_s|\omega) - 1 \approx -0.5(R_a - R_s)^2\omega^2c^{-2}, \qquad (50)$$

i.e., predict a red Doppler shift regardless of the relative positions of the light source and absorber on the rotor.

Kündig[39] also used the equivalence between acceleration and gravitation to discuss his experimental results and came to the same conclusion as Hay *et al.*[37] The conclusion from the theoretical analysis of both subsequent ultracentrifuge Doppler experiments[39,40] is therefore that an increase in light frequency was observed, that is, a shift in the opposite direction to that found with the Ives-Stilwell approach.[34,35,36] One can only speculate why this important distinction was not pointed out explicitly by these authors, but the suspicion is that they were dissuaded from doing so by the fact that the STR treatment of the transverse Doppler effect predicts unequivocally that a red shift will be observed in both cases. The very fact that Einstein's EP[38] was invoked to explain the results of the ultracentrifuge experiments implies that it cannot be simply a matter of perspective whether the absorber clock or that at the location of the light source is running slower.

Sherwin[41] clarified the interpretation of the experimental results shortly after Hay *et al.*[37] published their study by pointing out explicitly that there was no "ambiguity" as to which clock rate is slower. He went on to make the point that the fact that a blue shift is observed when the absorber is located at the rim of the rotor does not necessarily stand in contradiction to STR. This is because the absorber/detector is subject to high acceleration in the experiment and therefore does not satisfy the conditions for successful application of STR, namely that the "observer" be in uniform motion. However, this argument was much more plausible to make in 1960 than it was a decade later after the timing results for atomic clocks located on circumnavigating airplanes became available.[42] Despite the fact that these clocks were subject to minimal acceleration for the great majority of the time in flight, it was still found that it was completely unambiguous which of them had been running slower over the duration. This was in fact a very surprising result at the time of the atomic clock experiments, as the authors pointed out in their Introduction.

Taken together, the various transverse Doppler experiments support the conclusion in Chapter 4. B that the LT is not a valid space-time transformation. It therefore should not come as a surprise that its predictions regarding the ESP and the FLC are contradicted by the results of these investigations. The fact that the rotor experiments[37,39,40] can be satisfactorily interpreted in terms of Einstein's EP[38] certainly merits further discussion and this subject will be taken up subsequently when attention is turned to gravitational interactions. It also should not be forgotten that the transverse Doppler experiments verify the STR prediction of time dilation itself, but not in the symmetric form expected from the LT.

B. Meson Radiation Rates and Ranges

Shortly after Ives and Stilwell reported their results for the transverse Doppler effect, experiments were carried out with metastable particles[43,44] that also confirmed relativistic time dilation. Rossi *et al.* determined the average range before decay of cosmic-ray "mesotrons" and found that this quantity increased with the speed of the particles. They attributed this result to the increased lifetime of the particles due to their acceleration in the atmosphere, as expressed in eq. (47). The range L is defined to be equal to $v\tau$, the product of the particles' speed and their lifetime τ:

$$L = v\tau = \gamma(v)v\tau_0, \tag{51}$$

where τ_0 is the lifetime of the particles in the rest frame of the observer. For example, the lifetime of muons is 2.20×10^{-6} s. More precise laboratory measurements of this type have been made with focused particle beams from accelerators[44] and they are in complete agreement with eq. (45). Moreover, the latter experiments showed that the amount of time dilation is not affected by the degree of acceleration of the particles.[45] This result also stands in opposition to Sherwin's conjecture[41] that the ESP would begin to apply in place of asymmetric time dilation at the moment that the clock attains a state of pure translation. Instead, it would seem much more likely that rates would be unchanged by such a minimal alteration in the state of motion of the clocks.

The above lifetime measurements are in quantitative agreement with Einstein's asymmetric theory of time dilation.[5] They do not provide a direct test of the ESP since it is not possible to measure the lifetime of muons that are stationary in the laboratory from the vantage point of the rest frame of the accelerated particles, similarly as is the case for the Ives-Stilwell experiment.[34] However, they are also noteworthy for stimulating some additional discussion involving the RP. If one considers the decay process from the rest frame of the accelerated particles, it must be assumed there that the measured lifetime value is only τ_0, *i.e.* much shorter than the corresponding value measured in the laboratory. The fraction of particles $0 < X < 1$ that remains after they have reached the earth's surface must be the same for all observers, regardless of their state of motion. The only way this could be true for both the observer A moving with the particles and his counterpart B at rest on the earth's surface is if the distance to be travelled by the muons is γ times shorter for the former. Note that there is an implicit assumption in this argument that both observers agree on the speed v of the particles, as will be discussed subsequently.

The question is thus whether the fact that observer A moving with the muons measures a shorter distance to the earth constitutes a validation of the FLC. Although the answer in textbooks is invariably affirmative, the fact is that quite the opposite conclusion is required. The reason observer A measures a shorter elapsed time for his trip to the earth is because his proper clock runs slower than B's. The travel time is actually the same for both observers *but their respective units of time are different*. The numerical value of any measurement is inversely proportional to the unit in which it is expressed. The same holds true for distance measurements. The unit of distance is evidently γ times larger for A than for B. What A uses as his "meter stick" has a length of γm when measured in the standard units employed by B. The correct inference is therefore that the dimensions of objects increase at the same time, and by the same fraction, as their clocks slow down: time dilation is accompanied by isotropic length expansion rather than the FLC, which is exactly the same conclusion reached in Chapter 5. A. The fact that the range L of the meta-stable particles increases with their speed v relative to the laboratory is completely analogous to the increased wavelength λ observed in the experiments demonstrating the transverse Doppler effect. This result is also consistent with the argument of the clock riddle of Chapter 4. B. when clocks are used to measure distances.

The assumption that observers A and B agree on the value of the speed of the meta-stable particles requires some additional discussion, however. In the Ives-Stilwell experiment the assumption that both observers agree on the speed of light is obviously required by the light-speed postulate. According to the modern-day definition of the meter,[46] the length L of an object is to be determined by measuring the elapsed time T for a light pulse to pass between its two end-points and multiplying this result with c, which is defined to have a value of 2.99792458 ms^{-1}. When observers in two different rest frames S_1 and S_2 apply this definition, the following two general equations result for the length of the object: $L_1 = c_1 T_{1c}$ and $L_2 = c_2 T_{2c}$, where T_{1c} and T_{2c} are the respective elapsed times. For the time being, it is assumed that the corresponding two values of the light speed, c_1 and c_2, are not equal. An alternative means of measuring the object's length is to measure the elapsed times T_{1v} and T_{2v} for another object to move between the same two end-points. If one assumes that the corresponding measured speeds are v_1 and v_2, respectively, the following two equations result: $L_1 = v_1 T_{1v}$ and $L_2 = v_2 T_{2v}$. The key point is that the length of the object does not depend on whether a light pulse is used to measure it or some other quantity such as a car or a train. The ratio of the two measured lengths is then given by the following set of equations:

$$\frac{L_1}{L_2} = \frac{c_1 T_{1c}}{c_2 T_{2c}} = \frac{v_1 T_{1v}}{v_2 T_{2v}}. \tag{52}$$

On the other hand, the ratio of elapsed times measured by the two observers must be the same no matter what common event is considered, *i.e.* $T_{1c}/T_{2c} = T_{1v}/T_{2v}$. Therefore, if the light-speed postulate is now assumed, *i.e.* $c_1 = c_2$, it follows that

$$\frac{L_1}{L_2} = \frac{T_{1c}}{T_{2c}} = \frac{T_{1v}}{T_{2v}} = \frac{v_1 T_{1v}}{v_2 T_{2v}}. \tag{53}$$

The conclusion is therefore that $v_1 = v_2$, as was to be proved. In short, the light-speed postulate implies that two observers must also agree on the speed of any object moving between the two end-points.

In the case of the meta-stable particles, the two end-points are the location in which they are first accelerated by radiation and some other later position such as on the earth's surface. The equality of the measured relative velocities obtained by two observers leads to the following proportionality between their respective measured values of time and distance on a general basis:

$$\frac{L_1}{L_2} = \frac{T_1}{T_2}. \tag{54}$$

In short, the faster an observer's clock runs, the longer the distance L he will measure for the distance an object travels.

This in turn leads to the conclusion that the ratio of the units of length and time, *i.e.* the unit of velocity/speed, must be the same in all inertial systems. Strictly speaking, the latter conclusion is only valid if one adheres to an objective theory of the measurement process (PRM; Chapter 4. A). The ESP of STR actually excludes this possibility, whereas the argument given above is perfectly consistent with the PRM. In order to hold to the PRM, it is necessary to reject the FLC. The fact that both the wavelength of light and the range of meta-stable particles *increase* as clock rates slow because of time dilation constitutes strong experimental evidence that measurement is both objective and rational.

C. Rates of Atomic Clocks Onboard Circumnavigating Aircraft

In 1971 Hafele and Keating[42] carried out an experiment which further specified the theory of time dilation. Cesium beam clocks were flown in opposite directions around the earth. It was found that the clocks flying in an easterly direction lost time relative to their stationary counterparts at the point of origin, while those travelling in a westerly direction gained time. These results were obtained after correction for gravitational effects on the clock rates had been made. These authors found the following empirical formula for comparison of the respective elapsed times τ and τ_0 for the clock on the airplane and on the ground after the round-trip flight had been completed (R is the radius of the earth and Ω is its rotational speed at the Equator):

$$\tau - \tau_0 = - 0.5(2R\Omega v + v^2)c^{-2}\tau_0. \tag{55}$$

The sign convention is such that $v > 0$ if the airplane travelled in the easterly direction and $v < 0$ in the westerly direction. The above formula results if one assumes that the rate of a given clock is inversely proportional to $\gamma(v_{i0})$, where v_{i0} is its speed relative to "hypothetical coordinate clocks of an underlying non-rotating (inertial) space," *i.e.* the polar axis. Accordingly, the ratio of clock readings τ/τ_0 is given as:

$$\tau/\tau_0 = \gamma(R\Omega)/\gamma(R\Omega + v), \tag{56}$$

of which eq. (55) is an approximation to first order in $v_{i0}^2c^{-2}$.

The above quotation was clearly intended to argue that the LT was confirmed by the measured clock timings, but, as before with the rotor experiments in Chapter 5. A, no theoretical justification is actually achieved. The fact that the ESP is violated goes hand-in-hand with the realization that the LT is not internally consistent (Sects. IV. B-C). It also is worth noting that the polar axis is not an inertial system because the earth is in constant acceleration around the sun. The Hafele-Keating experiments are important for another reason, however. Eqs. (55-56) are seen to be identical in form to eq. (49) obtained by Hay *et al.* in their investigation of the transverse Doppler effect. A generalization of both results is obtained in the relation given below:

$$\tau_1\gamma(v_{10}) = \tau_2\gamma(v_{20}). \quad (57)$$

One has to substitute v_i^{-1} for τ_i in both cases for the rotor and use the rotor axis in the place of the polar axis from which to measure the corresponding v_{i0} speeds to be inserted in the γ factors. The same formula holds for Einstein's asymmetric theory of time dilation[5] for a clock moving with an accelerated electron, in which case the point where the force is applied serves as the reference system. The latter will henceforth be referred to in general as the objective rest system (ORS[47]), while eq. (52) merits the designation Universal Time-Dilation Law (UTDL). The earth's center of mass (ECM) can just as well be used as the ORS in place of the earth's non-rotating polar axis. One can speculate that the ECM plays a unique role in determining clock rates just as it does in gravitational interactions in the earth's environment. Accordingly, it can reasonably be expected that when a space capsule comes close enough to the moon to fall toward it instead of back to the ECM, the ORS for computing clock rates changes to the moon's center of mass. Moreover, when the rotor speed reaches a threshold value and gravity is overcome as a result, a transition to employing the rotor axis as ORS for determining clock rates also occurs.

D. Claims of Experimental Verification of the FLC

It has been shown in Chapter 4. B that the FLC does not satisfy the internal consistency requirements of a physical theory. On the other hand, it has been shown in the previous subsection that elapsed times satisfy the UTDL of eq. (57). Since the meter is defined as the distance travelled in c^{-1} s,[47] it is a simple matter to deduce an analogous relationship between the corresponding measured lengths of an object obtained by two observers in different inertial systems. One just has to multiply both sides of this equation with c and substitute the elapsed times for light to traverse the object's length λ_i (or corresponding distance between two fixed points in space) for the respective τ_i values:

$$\lambda_1\gamma(v_{10}) = \lambda_2\gamma(v_{20}). \tag{58}$$

This relationship is clearly also objective, unlike the FLC. It corresponds to *length expansion* (see Appendix A) accompanying time dilation since the observer with the slower clock, and therefore smaller τ_i value, will have a larger unit of distance in his rest frame because light travels farther in c^{-1} s than in the rest frame of the other observer. He therefore measures smaller λ_i values in accord with eq. (58). It always needs to be kept in mind that the numerical value of a measurement is inversely proportional to the unit in which it is expressed. Thus if $v_{10} > v_{20}$, then $\lambda_1 < \lambda_2$ and $\tau_1 < \tau_2$, because both the period of a proper clock and the length of a meter are greater in S_1 than in S_2.

The FLC is predicted on the basis of the LT and many authors have argued that it is theoretically impossible to obtain verification of this effect because of the supposed requirement of simultaneous measurement of both ends of an object while it is moving relative to the observer. The advent of GPS countered that position but, as has been shown above, this development actually speaks against the FLC's occurrence in nature. As will be discussed in Chapter 5. E, other claims for the validity of the FLC have been based on assumptions that rely on the LT for their justification.

In recent times there has nonetheless been a claim for the validity of the FLC which is based on experiments studying the Josephson effect. Laub *et al.*[48] found that magnetic flux quanta in Josephson tunnel junctions undergo Lorentz contraction. In annular junctions pairs of vortices and anti-vortices are created that move in opposite directions and "collide" with each other. The collision region was visualized in their experiments with the aid of low-temperature electron microscopy. It was assumed that the length of the collision region is proportional to the length of the vortices. The collision region was found to contract with increasing vortex velocity. A decrease in the length of this region with increasing voltage, which is proportional to the vortex speed, is clearly visible in the authors' Fig. 4.[48] The experiments are not able to quantitatively verify that the collision length varies in direct proportion to γ^{-1} (v), however, as prescribed by the FLC.

In analyzing the Laub *et al.* data, it is important to keep in mind that the length of the collision region is a measure of the density of electrons moving in opposite directions through the Josephson apparatus. Consequently, *there is another reason for the decrease in length that has nothing whatsoever to do with Lorentz contraction.* The distribution of such large numbers of electrons is characterized by a de Broglie wavelength λ[49] which is inversely proportional to their momentum p in the laboratory ($\lambda p = h$, Planck's constant). That this effect is distinct from Lorentz contraction is seen from the fact that the velocity dependence of the FLC is the same for all types of particles, whereas the de Broglie wavelength contraction will be far greater for collections of protons than for electrons moving at the same speed. Without doing more quantitative studies, it is impossible to know which of the two effects is actually responsible for the observed contraction, but there is no reason to discount at least a partial effect caused by de Broglie wave-particle duality. [49] This possibility has been ignored in the discussion given by Laub *et al.*[48] Similar remarks hold for more recent experiments[50] demonstrating an increase in the frequency of synchrotron radiation when electrons are accelerated. Increased ionization is also expected when the de Broglie wavelength of a distribution of nuclei is shortened by virtue of an increase in their speed relative to the laboratory.

E. Verifications of the VT

The theoretical predictions based on the VT have been confirmed in all known cases, but there is confusion in the scientific literature regarding their significance. The foremost example is the Michelson-Morley experiment itself.[3] The fact that only a null result is observed for the difference between the elapsed times for light to traverse both arms of the interferometer is usually interpreted to be a verification of the FLC. In actuality, the null result is direct confirmation that the speed of light is independent of the orbital speed of the earth, which is a fundamental consequence of the VT alone. The conclusion that the Michelson-Morley experiment confirms the FLC is based on an assumption, namely that there is a one-to-correspondence between the VT and the LT, but this is incorrect. As shown in Chapter 2. B, the VT can be derived from the GLT of eq. (15) for any value of the normalization factor ε. In other words, there is a many-to-one correspondence involved. Thus, verifying the VT in no way guarantees that any one of an infinite number of space-time transformations consistent with the definition of the GLT is also confirmed. The same *caveat* holds for the Kennedy-Thorndike experiment[51] with unequal arms in its apparatus. In that case the experiment is claimed to verify both the FLC and time dilation because they in turn are both derived from the LT. For this conclusion to be valid, it would be necessary to show that only the LT can satisfy the two postulates of relativity. The fact is that the arguments of Chapter 4. B (clock riddle) show that the LT is excluded as a valid transformation because it does not satisfy the basic self-consistency condition required of any valid physical theory. In summary, both the Michelson-Morley and Kennedy-Thorndike experiments only serve as confirmatory evidence for the VT. They tell us nothing definitive about whether the LT, and therefore the FLC and time dilation, are themselves supported by these experiments.

Similar remarks hold for the aberration of starlight at the zenith. The relation in eq. (37), which is of fundamental importance in astronomical observations, can be derived on the basis of the VT alone. Yet it is invariably referred to as one of the prime examples illustrating the validity of the LT.[4] Any choice of ε in eq. (15) leads to the same result as the LT, and so there is no reason to regard the aberration of starlight as a unique confirmation for it.

The Fresnel light-drag experiment is another clear confirmation of the VT. Its derivation by von Laue in 1907[16] is based directly on the VT. He assumed that the speed of light in a condensed medium of refractive index n is c/n. Substitution of this value in the VT leads directly to eq. (32) and gives the empirical result for the velocity dependence of the observed light speed in the laboratory as a function of the speed of the moving medium. The VT also guarantees that the speed of a particle never exceeds the value of c for any observer regardless of his state of motion, a fact which has been confirmed many times in the laboratory. The corresponding theoretical result is expressed in eq. (38) and is again derived exclusively on the basis of the VT.

The many successful applications of the VT emphasize that a valid theory of relativity must be consistent with it. However, they also show that it is sufficient for the theory to be consistent with the GLT of eq. (15), since this guarantees satisfaction of the

VT. In the next chapter, consideration will be given to how a different assumption than Einstein[5] used to derive the LT from the GLT can remove the contradiction revealed by the clock riddle and produce a new space-time transformation that is consistent with all experimental data as yet obtained.

Chapter 6

Alternative Lorentz Transformation (ALT)

The discussion in the previous chapters has underlined three key points. In Chapter 4 it has been shown that the LT lacks internal consistency (clock riddle[32]) and is therefore invalid. The VT on the other hand has been confirmed by numerous experimental investigations (see Chapter 5. E) and therefore should be a central fixture in any attempt to construct an amended version of relativity theory. Finally, it has been found in Chapter 5. C that eq. (57) successfully predicts the results of all previous experiments involving time dilation. This relationship, which has been referred to as the UTDL, rejects the ESP of STR, and instead assumes that the clock readings of observers in relative motion satisfy an asymmetry principle. On the basis of the above conclusions it is possible to derive a new relativistic space-time transformation that removes the contradiction inherent in the LT transformation, but is still consistent with both the UTDL and the VT.

A. Derivation of the ALT

The lack of symmetry in the time-dilation experiments does not contradict in any way the light-speed postulate employed by Einstein to derive the LT.[5] What it suggests instead is that his assumption regarding the degree of freedom in the general form of the Lorentz transformation, the GLT of eq. (15), needs to be reexamined.[33,52,53] There is no clear justification for his assertion (see Chapter 2. C) that ε can only be a function of the relative speed v of the two rest frames and therefore for his conclusion that it can only have a constant value of unity. Given the experimental facts it makes more sense to insist that ε be chosen so as to maintain consistency with the UTDL empirical formula of eq. (57). This can easily be brought into a form which is consistent with the notation in eq. (15) by assuming the following proportionality:

$$\Delta t' = \Delta t Q^{-1}, \tag{59}$$

where Q is a ratio of the two elapsed times in eq. (57) and/or the corresponding ratio of the two $\gamma(v_{i0})$ factors, *i.e.*, $Q = \gamma(v_0')/\gamma(v_0)$. This relation allows one to obtain a different value for the undetermined function ε by comparing with the first relation in eq. (15):

$$\Delta t' = \gamma\varepsilon(\Delta t - v\Delta x c^{-2}) = \Delta t Q^{-1}. \tag{60}$$

Solving for ε yields:

$$\varepsilon = [(1 - v\Delta x \Delta t^{-1} c^{-2})\gamma Q]^{-1} = [(1 - v u_x c^{-2})\gamma Q]^{-1} = \eta(\gamma Q)^{-1}, \tag{61}$$

where η is the same function that appears in each of the VT relations in eq. (29). Inserting the above value for ε in the GLT of eq. (15) therefore produces a new space-time transformation, hereafter referred to as the alternative Lorentz transformation (ALT[33,52,53]), that is consistent with Einstein's light-speed postulate as well as with the UTDL, namely:

$$\begin{aligned} \Delta t' &= Q^{-1}\,\Delta t \\ \Delta x' &= \eta Q^{-1}(\Delta x - v\Delta t) \\ \Delta y' &= \eta(\gamma Q)^{-1}\,\Delta y \\ \Delta z' &= \eta(\gamma Q)^{-1}\,\Delta z, \end{aligned} \tag{62}$$

whereby the first of these equations is identical to eq. (59). Note that the proportionality factor Q also appears in the spatial coordinate relations in eq. (62). The above result is clearly necessary[53] in order to ensure that the new transformation is consistent with the VT of eq. (29). However, it also has important consequences for the other of Einstein's postulates of relativity, the RP.

B. Lorentz Invariance

One of the most attractive features of STR is its Lorentz invariance condition. It follows directly from the GLT by squaring and adding the four relations in eq. (15):

$$\Delta x'^2 + \Delta y'^2 + \Delta z'^2 - c^2\Delta t'^2 = \varepsilon^2(\Delta x^2 + \Delta y^2 + \Delta z^2 - c^2\Delta t^2). \tag{63}$$

Lorentz invariance is the special case for this equation when the value of $\varepsilon = 1$ is substituted therein [see eq. (11)]. When both S and S′ are inertial systems, the RP demands that respective observers who are stationary in either of these rest frames find the same laws of physics to be valid. That implies a definite symmetry that is far from obvious from the general form of eq. (63). As noted above, Einstein derived the LT from eq. (15) by claiming that $\varepsilon = 1$ is the only acceptable value for this quantity.[5] It is clear that this choice produces a symmetric appearance for eq. (63) as well as the condition of Lorentz invariance.

When any other value for ε is assumed, such as that used to derive the ALT in eq. (62), it is critical to show that the RP is also satisfied by this choice. One might try to circumvent this question by insisting that the latter transformation is only valid for non-inertial systems, but such a position does not stand up to further scrutiny for the reason that an accelerated clock can assume a state of uniform translation at any time. When this happens, there is no evidence to indicate that its clock rate changes as a consequence. On the contrary, the experimental data for muon decays (see Chapter 5.B) in storage rings[43,44,45] indicate that only the speed of the particles affects their lifetime, and not the level of acceleration to which they are subjected.

The question thus arises as to whether the choice of $\varepsilon = \eta (\gamma Q)^{-1}$ that leads to the ALT is also consistent with the RP. Substitution in eq. (63) gives the following alternative condition of invariance:

$$\Delta x'^2 + \Delta y'^2 + \Delta z'^2 - c^2\Delta t'^2 = \eta^2(\gamma Q)^{-2} (\Delta x^2 + \Delta y^2 + \Delta z^2 - c^2\Delta t^2). \quad (64)$$

To satisfy the RP, it is necessary for the inverse of eq. (64) to have the same form from the vantage point of the observer in S, i.e

$$\Delta x^2 + \Delta y^2 + \Delta z^2 - c^2\Delta t^2 = \eta'^2(\gamma Q')^{-2}(\Delta x'^2 + \Delta y'^2 + \Delta z'^2 - c^2\Delta t'^2). \quad (65)$$

In this equation η' must be obtained from $\eta = (1 - vu_x c^{-2})^{-1}$ in the standard way by exchanging corresponding primed and unprimed values and changing v to $-v$, *i.e.* $\eta' = (1 + vu_x'c^{-2})^{-1} = (1 + v\Delta x'\Delta t'^{-1}c^{-2})^{-1}$. The value of γ remains the same because it is a function of v^2. The value of $Q' = Q^{-1}$ is fixed by forming the inverse of eq. (62), *i.e.* $\Delta t = Q'^{-1} \Delta t' = Q \Delta t'$.

There is another way to invert eq. (64), namely to divide both sides by $\varepsilon^2 = \eta^2(\gamma Q)^{-2}$, with the result:

$$\Delta x^2 + \Delta y^2 + \Delta z^2 - c^2\Delta t^2 = \eta^{-2}(\gamma Q)^2(\Delta x'^2 + \Delta y'^2 + \Delta z'^2 - c^2\Delta t'^2). \quad (66)$$

Both equations must be equivalent in order to satisfy the RP, hence $\eta^{-2}(\gamma Q)^2$ in eq. (66) must be equal to $\eta'^2(\gamma Q')^{-2}$ in eq. (65). The primed variables in the definition of η' can be eliminated by using the ALT/VT, whereupon the following identity is obtained (see Appendix B):

$$\eta\,\eta' = \gamma^2. \tag{67}$$

Consequently, eqs. (65) and (66) are seen to be equivalent since $Q' = Q^{-1}$. The choice of $\varepsilon = \eta\,(\gamma Q)^{-1}$ in the GLT of eq. (15) to define the ALT therefore satisfies both of the relativity postulates just as well as Einstein's $\varepsilon = 1$ value does for the original LT. Yet, unlike the LT, it is consistent with the asymmetry that is always observed in two-way time-dilation tests.

C. Uniform Scaling and the Amended Relativity Principle

There is another topic that needs careful attention when proposing a new version of the Lorentz transformation to remove the experimental contradictions that result from reliance on Einstein's LT. The discussion in the preceding paragraph notwithstanding, it is clearly necessary to show explicitly how a different space-time transformation guarantees satisfaction of the RP.

According to the ALT of eq. (62), elapsed times measured in different inertial systems always differ by a definite proportionality factor Q. This guarantees that time dilation is asymmetric, in agreement with all experiments. Unlike the case for the LT, however, timings are not subject to space-time mixing, and this fact greatly simplifies matters as far as satisfaction of the RP is concerned. It allows one to look upon Q in the above equation as a "conversion factor" between the respective units of time in S and S′, respectively.[54] The laws of physics are ordinary equations that are independent of the units chosen for the various quantities that appear therein. For example, the laws are not affected in any material way when the unit of length is changed from meters to yards or the unit of force from Newtons to pounds. All changes in the numerical values of the quantities contained in the laws that are caused by such changes in units are simply cancelled out in the overall equations so that the laws themselves are unaffected. This is exactly the situation when the ALT is used to determine the conversion factor between the different units of time in the two inertial systems. The RP is therefore satisfied in all cases, as Einstein's first postulate of relativity demands.

Moreover, it is also known that the unit of speed/velocity must be the same in all inertial systems because otherwise it is impossible, as required by Einstein's second postulate, to explain how the speed of light has the same value in each of them (see Chapter 5.B and Appendix C). Therefore, no conversion factor for velocities is required to change from one inertial system to another. The constancy of the unit of velocity has another important consequence, however. The only way the values of speeds can be the same for observers in different inertial systems, even though their respective timing measurements differ, is for their length measurements to differ by exactly the same factor. Since the unit of time varies with Q according to the ALT while the unit of speed does not change at all, it therefore follows that the unit of distance must vary in exactly the same way as that of time. One of the consequences of this state of affairs is that length measurements are also asymmetric, as indicated by eq. (58). There is never any ambiguity as to which of two lengths is greater, just as there is no question about which clock rate is slower. The RP is automatically satisfied for distances because any law that is valid in one initial system which involves distances must also be valid in any other inertial system.

The above result for distance variations is quite different than what Einstein predicted using the LT.[5] He concluded with the FLC that lengths contract by the same fraction that clocks slow down and that the effect is anisotropic, *i.e.* they decrease along the parallel direction of motion relative to the observer but they do not change at all in perpendicular directions. Consistent with Lorentz invariance, the effect should be symmetric for two observers, the same characteristic as he found for time dilation. The Ives-Stilwell transverse Doppler experiments provide a straightforward test for these predictions.[34] It was found (see Chapter 5. A) that the wavelength of light emitted in a perpendicular direction from a moving source *increases* by a factor of $\gamma(v)$. Moreover, the form of the general relativistic Doppler equation indicates that the same increase in wavelength occurs in all directions once the non-relativistic contribution due to relative motion of the source and detector is accounted for.[55] In accordance with the RP, an observer co-moving with the light source does not notice any change in the standard wavelength of the light source. This conclusion indicates that the dimensions of the diffraction grating used in the local measurement have changed by exactly the same fraction in all directions as the wavelength of the standard light source. The result is perfectly consistent with the isotropic length expansion expected from the asymmetric theory of time dilation, but it stands in direct contradiction to the FLC predicted on the basis of the LT.

The general treatment of changes in clock rates and dimensions of objects discussed above can be concisely summarized in terms of the *uniform scaling* of physical properties that is caused by application of force. The conversion factor for energy and inertial mass is the same as for time and distance, *i.e.* the value of Q in the ALT connecting the two inertial systems.

The above considerations indicate that the conventional statement of the RP needs to be amended as follows:

> The laws of physics are the same in all inertial systems but the units in which they are expressed can and do vary in a systematic manner from one rest frame to another.

Physical laws are mathematical equations. They must retain their validity when the system of units is changed, whether this involves simply expressing results in a different set, *or because the standards used to define these quantities have undergone changes themselves*. Experiment tells us that the clock rates, lengths and inertial masses of standard objects change when they are accelerated, and so this must be taken into account when comparing measured values obtained for the same quantity by observers who are stationary in different inertial systems.

The constant Q in the ALT is a *conversion factor* between these units and therefore must appear explicitly in the space-time transformation relating measured values in the two participating inertial systems. In a companion article,[56] it has been shown that the conversion factors for the units of all physical properties are always powers of Q. Moreover, a different constant determines the corresponding conversion factors required to account for the effects of changes in gravitational potential, as discussed below (Chapter 10. B) when attention turns to the general topic of gravitation. Together these values allow one to quantitatively predict how the measurements of the same property will differ for observers located at different positions in the gravitational field and/or in different states of motion.

D. Derivation of the Relativistic Doppler Relations

One of the most often discussed results of STR is the relativistic Doppler equation. As has been mentioned in Chapter 3. B, Einstein's prediction of the transverse Doppler effect[5] was new to physicists at the time, and was first verified in the experiments of Ives and Stilwell[34] 33 years later. According to eq. (26), an observer should always measure a decrease in frequency ν_r relative to the standard value ν_e for the stationary light source. The latter relation is derived from the general angle-dependent Doppler equation which is derived on the basis of the LT,[36,57] where ν' is the measured frequency in the rest frame of the light source:

$$\nu' = \gamma\nu\,(1 - vc^{-1}\cos\Theta). \tag{68}$$

The direction of propagation of the light waves relative to the velocity vector of S and S′ is expressed in terms of the polar angle Θ. The transverse case occurs for $\Theta = \pi/2$, which corresponds to the result in eq. (26) except for the difference in notation.

Since the LT has been shown to be invalid, it is clearly desirable to obtain a different derivation that leads to the correct experimental predictions that rests on a solid theoretical foundation. Before doing this, however, it is interesting to look at the derivation based on the LT. It starts out with the assumption that the phase $\Phi = \mathbf{k}\cdot\mathbf{r} - \omega t$ of a wave is invariant to a Lorentz transformation, where $\mathbf{k}$ is the wave vector with absolute value $k = 2\pi/\lambda$ and $\omega = 2\pi\nu$. This argument is based on the plausible assertion that the number of waves reaching a certain point of space over a definite period of time must be the same for all observers. It is assumed that the wave originates in S, so the next step is to look at it from another inertial frame S′ moving with constant velocity $\mathbf{v}$ along the x-axis relative to S. The LT then leads to the following relationships between the two sets of k components and ω given below:

$$\begin{aligned} \omega' c^{-1} &= \gamma(\omega c^{-1} - v k_x c^{-1}) \\ k_x' &= \gamma(k_x - v\omega c^{-2}) \\ k_y' &= k_y \\ k_z' &= k_z. \end{aligned} \tag{69}$$

Converting the first of these relations to polar coordinates, with $k_x = k\cos\Theta$, leads to:

$$\omega' c^{-1} = \gamma(\omega c^{-1} - vkc^{-1}\cos\Theta). \tag{70}$$

Since these are electromagnetic waves, $k = \omega c^{-1}$, and therefore eq. (70) becomes eq. (68) upon substitution of $\omega = 2\pi\nu$ and $\omega' = 2\pi\nu'$.

There are two problems with this derivation. First, note that the observer in this example is at rest in S′, whereas the light source is stationary in S. Thus, the frequency at the light source (ν_e) in eq. (26) corresponds to ν in eq. (68) and ν_r corresponds to ν'. Substitution gives $\nu_r = \gamma\nu_e (1 - vc^{-1}\cos\Theta)$. Setting $\Theta = \pi/2$ does not lead to eq. (26) but rather to $\nu_r = \gamma\nu_e$. However, it can be argued that Θ is the polar angle in S where the light source is located, and thus one should compare with the inverse of eq. (70) which is obtained in the usual way by exchanging the primed and unprimed indices and changing v to –v. The corresponding relation is then:

$$\omega c^{-1} = \gamma(\omega' c^{-1} + vk' c^{-1}\cos\Theta'). \qquad (71)$$

Making the same substitution with $\Theta' = \pi/2$ then gives $\nu_e = \gamma\nu_r$, in agreement with eq. (26) for the transverse case, but it fails to give the observed dependence on polar angle, which will be discussed subsequently.

The other problem has to do with the derivation of the LT itself. In Chapter 2. B it was stipulated that the coordinate axes in S and S′ coincide at $t = t' = 0$. Not all pairs of inertial systems satisfy this requirement, so the claim that the measured frequency and wave vectors must have the relationships given by eq. (69) is incorrect.

What is needed is a different derivation of the Doppler relations which starts out with the experimental fact that the rates of proper clocks in different inertial systems can occur in any conceivable ratio. This is clear from the UTDL in eq. (57). The simplest way to accomplish this goal is to first compute the relationship between wavelengths λ and λ' measured in the two inertial systems.[58] Assume that the light source is at rest in S′ and the wavelength measured there is λ'. The corresponding wavelength λ measured in S is therefore $Q\lambda'$, where Q is the conversion factor defined in the ALT of eq. (62), if the source is moving transversely to the observer. The general relation must take into account that the distance between peaks of the radiation depends on the polar angle of approach Θ of the light source. If the period of the radiation is $T' = \lambda' c^{-1}$ in S′, then

$$\lambda = Q\,\lambda' - v\,QT'\cos\Theta = Q\lambda'\,(1 - vc^{-1}\cos\Theta), \qquad (72)$$

where $\Theta = 0$ corresponds to a head-on approach of the light source to the observer. Note that the factor multiplying $Q\lambda'$ is non-relativistic in nature. It makes the standard correction required because the light source in either moving into or out of the wave pattern. The conversion factor Q is a direct measure of both time dilation and also length expansion in S′ relative to S. Unlike eq. (26), eq. (72) corresponds to asymmetric time dilation, whereby the angle Θ is always measured relative to the velocity **v** of the two inertial systems from the vantage point of the observer in S. The inverse of eq. (72) for radiation coming from a stationary light source in S that is measured by an observer who is at rest in S′ is thus:

$$\lambda' = Q'\lambda\,(1 - vc^{-1}\cos\Theta'), \tag{73}$$

where $\Theta' = 0$ now corresponds to a head-on approach of the light source to the observer and $Q' = Q^{-1}$. The usual rule for inverting equations does not apply for the factor in parentheses because, as already mentioned, it is non-relativistic in nature. Both the latter two equations are consistent with the ALT, but cannot be derived from the LT even though they correspond to the empirically observed results in all cases.

E. Replacing the LT by the ALT

The discussion above has shown that a fundamental change in relativity theory needs to be made in order to counter the lack of internal consistency in Einstein's original version of STR.[5] The starting point is to eliminate the LT as the cornerstone of the theory. There are many consequences for doing this that need to be clearly understood. For example, the condition of Lorentz invariance must be discarded because the corresponding value of $\varepsilon = 1$ in eq. (63) is no longer tenable. The ALT value given in eq. (61) satisfies the requirement that $\varepsilon\varepsilon' = 1$ and guarantees conformance to the RP, as shown in Chapter 6. B. The new invariance condition is given by eq. (64) and the inverse relation in eq. (65).

Eliminating the LT also means that the concept of space-time mixing loses its primary justification. Eq. (42) of the LT indicates that the elapsed time $\Delta t'$ measured by one observer depends not only on the corresponding elapsed time Δt for the other observer but also the distance Δx separating the two events. This characteristic of STR has led to the term "spacetime, " which is widely used in the scientific literature. It does away with the Newtonian separation of space and time, which holds that the three spatial coordinates are fundamentally different than the time coordinate. Spacetime is four-dimensional and cannot be separated into the one-dimensional time and three-dimensional spaces that seem so intuitively reasonable in our everyday existence. Without the LT, and particularly with the ALT, the revised theory holds that Δt and $\Delta t'$ are strictly proportional to one another. According to eq. (59) of the ALT, the ratio of these two quantities simply varies with the relative speed of the two inertial systems, in agreement with the empirical timing results obtained by Hafele and Keating[42] and also with the underlying principles of the GPS technology.[59] It does not matter in any way where the two clocks are located in normal three-dimensional space.

Space-time mixing in the LT led to speculation[4] of *remote non-simultaneity* of events. According to eq. (42), if two events are at different locations in S ($\Delta x \neq 0$) it follows that if they occur simultaneously for an observer who is stationary in that rest frame ($\Delta t = 0$), they will not be simultaneous in any other inertial system ($\Delta t \neq 0$). This idea was revolutionary in 1905, but it quickly became doctrine as belief in STR and the LT established itself in the coming years. Popular magazines published fanciful articles in which, for example, two businessmen at different locations were unable to sign a contract at exactly the same time, therefore jeopardizing their proposed deal. Because of the proportional relationship in eq. (59) of the ALT, one must assume on the contrary that Δt and $\Delta t'$ must always vanish together, returning to the long-standing belief in remote simultaneity that before the 20th century was widely accepted as fact by physicists.

Another prediction of the LT that no longer needs to be taken seriously is that time reversal can occur between inertial systems moving with relative speed $v > c$. Examination of eq. (42) shows that the ratio of Δt and $\Delta t'$ can be negative under this condition, which would mean that the time-order of events is opposite for observers in the two rest frames. This eventuality has been widely considered to violate common sense even in the context of high-energy physics, leading to the conclusion that no object can move at faster-than-c speed. For example, in 2011 the OPERA collaboration[60] reported evidence that neutrinos move with $v > c$, and thereupon encountered strong resistance from the physics community on the basis of the above theoretical grounds. The controversy on this issue dates back to much earlier times, however. It was pointed out that the group refractive index n_g is less than unity for condensed media in wavelength regions of *anomalous dispersion* (near absorption lines).[61] Since the speed of light in condensed media was found experimentally[20] to be equal to cn_g^{-1} for normally dispersive liquids (with $n_g > 1$), it was argued that light speeds exceeding the value in free space might be possible in media with $n_g < 1$. In 1993 new experimental evidence[62,63] emerged that appeared to demonstrate unequivocally that $u > c$ light speeds were indeed attainable in media with $n_g < 1$. However, even these results were not sufficient to dispel the general reluctance on the part of the physics community to accept as fact that single photons can indeed travel with faster-than-c speeds under the above conditions.[63,64] The ALT of eq. (62) by contrast is non-committal about the maximum speed attainable. Einstein concluded in his original work[5] that particles with non-zero rest mass may not travel with $v > c$, but thereby left open the possibility that photons can, so there are in fact no incontrovertible theoretical grounds for doubting the validity of the above experimental data.[62,63] More discussion of this issue will be given in Chapter 14.

The ALT also removes any grounds for belief in either the FLC or the ESP, as has already been discussed in Chapter 5. The arguments for these predictions based on the LT are super-ceded by the UTDL of eq. (57) for elapsed times and the corresponding relation for measured distances given in eq. (58).

Finally, perhaps the most sweeping conclusion of the LT and STR is the claim that the laws of physics must be invariant to this transformation. Realization that the LT is invalid removes any grounds for adhering to this principle. Instead, one can test the validity of empirical laws on the basis of the amended version of the RP (see Chapter 6. C). One merely has to show that a uniform scaling of the units in which the laws are expressed leaves the corresponding equations invariant. This point will be discussed in detail in the following section when attention is turned to relativistic effects for quantities other than space and time.

The above summary shows that there have been many attempts to either dismiss experiments that run contrary to the predictions of the LT or to give a less-than-definitive argument for one that supposedly verifies it. This situation underscores the importance of dealing with the lack of internal consistency in the underlying logical framework of this space-time transformation. The clock riddle discussed in Chapter 4. B is based solely on the well-known conclusions of the LT, namely the FLC and time dilation applied in conjunction with Einstein's second postulate. Failure to prove that there is something wrong with either the premises of the clock riddle or the manner in which logic has been applied to obtain its conclusions constitutes indisputable proof that the LT is not credible. Once this consequence has been accepted, there is no difficulty in showing that the VT and the ALT satisfactorily account for any experimental result that has heretofore been claimed to be irrefutable evidence for the validity of the LT.

Chapter 7

Energy, Momentum and Inertial Mass

The foregoing discussion has concentrated exclusively on space and time relationships. Many of the most significant results of Einstein's STR[5] involve other quantities, however, including his famous mass-energy equivalence relation: $E = mc^2$. The present chapter describes how the light-speed postulate can be applied to obtain the relativistic analogue of the classical definition of momentum. It will be shown that Newton's Second Law of Kinematics provides an excellent starting point to derive the relativistic transformation of energy and momentum. Once again it is assumed that all interactions take place at the same gravitational potential, exactly as Einstein assumed in his original work.[5]

A. Einstein's Derivation of the Mass-Energy Equivalence Relation

The emphasis in developing a relativistic theory to explain the results of experiments carried out in the latter half of the 19th century was on the derivation of a space-time transformation that leaves Maxwell's equations invariant. Einstein's analysis[5] led him to conclude that there were two kinds of inertial mass: longitudinal m_L and transverse m_T. The expression he obtained for the former is

$$m_L = \mu\gamma^3, \qquad (74)$$

where μ is the rest mass of the particle, v is its speed relative to the observer and $\gamma = (1 - v^2c^{-2})^{-0.5}$, as defined in eq. (14). When the particle is subjected to a force F, electromagnetic or otherwise, its kinetic energy W is equal to ∫Fdx, where dx is the differential distance the particle moves as a result of the applied force. Einstein then stated without further discussion[5] that on the basis of eq. (68),

$$W = \int m_L \, vdv = \mu \int \gamma^3 vdv. \quad (75)$$

One can only speculate how he arrived at this conclusion, but it would at least appear that he was extending the non-relativistic result for kinetic energy W_{nr}:

$$W_{nr} = \int Fdx = \int (dp/dt)dx = \mu \int vdv, \quad (76)$$

i.e., with the non-relativistic momentum p = μv and v = dx/dt. In any event, Einstein continued his argument by computing the definite integral on the right-hand side of eq. (75) from 0 to v ($c^2 d\gamma = \gamma^3 vdv$), hence:

$$W = \mu c^2 \gamma \Big|_0^v = (\gamma - 1)\mu c^2 = mc^2 - \mu c^2. \quad (77)$$

The above equation contains the definition of relativistic (inertial) mass m as

$$m = \gamma\mu, \quad (78)$$

which is seen to bear a close resemblance to the time proportionality relation of the ALT in eqs. (59,62). The kinetic energy W is defined as the increase in total energy E of a system by virtue of its motion relative to its initial position prior to being accelerated, *i.e.* W = E – E′, where E′ is the system's rest energy. It therefore follows by making the definition, $E' = \mu c^2$, in eq. (77) that

$$E = \gamma E', \quad (79)$$

from which Einstein's mass-energy equivalence relation follows:

$$E = mc^2 = \gamma\mu c^2. \quad (80)$$

Although the above equations are suitably consistent with one another, their validity can be questioned on the basis of its reliance on eq. (75) to define W. Planck suggested[65] that a more theoretically consistent derivation can be obtained by applying Newton's Second Law, **F** = d**p**/dt, and assuming that the momentum **p** is defined as

$$\mathbf{p} = m\mathbf{v} = \gamma\mu\mathbf{v}. \quad (81)$$

Einstein agreed and revised his derivation as follows:[38]

$$W = \int Fdx = \int (dp/dt)dx = \int vdp = \mu \int vd(\gamma v) = \mu \int \gamma^3 vdv, \quad (82)$$

in place of eq. (75). Thus, Einstein's results in eqs. (77-80) are shown to follow directly from Newton's Second Law and the definition of relativistic momentum given in eq. (81), whereby eq. (78) for relativistic mass is simply shown to be consistent with the latter definition in this approach.

B. Derivation Based on Hamilton's Canonical Equations

The result in the previous section suggests another way to derive the above equations, namely by combining Hamilton's canonical equations, specifically

$$dE = vdp, \tag{83}$$

with the light-speed postulate. The latter is closely related to Newton's Second Law, as is evident from eq. (82) after equating dE to dW. One can look upon this equation as a definition of the speed v of an accelerated object as the rate of change of its energy E with respect to the change in its momentum p_x in the direction of motion. In analogy to the Galilean transformation (GT; see Chapter 2. A), it is possible formally to convert eq. (83) to a relation between differential energies observed in two rest frames S and S′:

$$dE' = dE - v\,dp_x. \tag{84}$$

The light-speed postulate can then be used to insure that $dE/dp_x = dE'/dp_x' = c$ for the special case when the particle is a photon. In analogy to the derivation of the GLT of eq. (15) in Chapter 2. B, one then obtains the following general relativistic energy-momentum transformation:

$$\begin{aligned} dE' &= \varepsilon\gamma(dE - v\,dp_x) \\ dp_x' &= \varepsilon\gamma(dp_x - vc^{-2}\,dE) \\ dp_y' &= \varepsilon dp_y \\ dp_z' &= \varepsilon dp_z' \end{aligned} \tag{85}$$

The quantity ε needs to be specified by imposing an additional condition on the transformation, similarly as for the GLT but quite independently because of the difference in variables involved. A value of ε = 1 will be assumed in the following discussion for purposes of making connection with the results obtained in Chapter 7. A.

There is an important special case for eq. (85), namely when the object is at rest with respect to the origin of the acceleration (**dp**′ = 0). For this purpose, it is useful to first work with the inverse transformation:

$$dE = \gamma(dE' + v\, dp_x')$$

$$dp_x = \gamma[dp_x' + vc^{-2}dE'] \qquad (86)$$

$$dp_y = dp_y'$$

$$dp_z = dp_z'$$

The differential quantities in these relations can be looked upon as infinitesimal parts of macroscopic quantities. With this interpretation, integration of eq. (86) leads directly to Einstein's eq. (79) with $dp_x' = 0$. The same condition leads to the mass-energy equivalence relation. This is seen from the second relation in eq. (86) and recalling that inertial mass m is defined as the ratio of momentum (p_x) to speed v:

$$dp_x = m\, v = dEc^{-2}\, v = \gamma dE'c^{-2}\, v = \gamma\mu\, v, \qquad (87)$$

i.e. $dE = mc^2$. It is also consistent with the definition of relativistic momentum in eq. (81). This result also shows that the rest energy of an object is generally not zero since it must be proportional to the rest (or proper) mass $\mu = m\ (v = 0) = m'$, which is the value measured *in situ* by all observers:

$$m = \gamma dE'c^{-2} = \gamma m' = \gamma\mu, \qquad (88)$$

in accordance with eq. (78). Finally, the value of ε in the general transformation of eqs. (85-86) can be determined uniquely on the basis of the non-relativistic kinetic energy

$$W_{nr} = uv^2/2. \qquad (89)$$

It is seen to be unity from eq. (77) since in the limit of small velocities,

$$W = W_{nr} = \varepsilon(\gamma - 1)\mu c^2 \approx \varepsilon\mu c^2\, (v^2c^{-2}/2) = \varepsilon\, W_{nr}. \qquad (90)$$

The E-p transformation in eqs. (85-86) with ε = 1 leads to the following invariance relation upon squaring and adding the four equations:

$$dE'^2 - c^2\, (dp_x'^2 + dp_y'^2 + dp_z'^2) = dE^2 - c^2\, (dp_x^2 + dp_y^2 + dp_z^2). \qquad (91)$$

It has the symmetric appearance associated with space-time Lorentz invariance in eq. (11), which is the effect of employing the value of $\varepsilon = 1$ in the transformation, but in this case no contradiction results from this choice. For the special case of $dp' = 0$, the invariance relation becomes (after integration):

$$E'^2 = \mu^2c^4 = E^2 - p^2c^2. \tag{92}$$

This equation has been verified to high accuracy in laboratory experiments, thus reinforcing the belief in the above derivations. Because $\varepsilon = 1$ in the E-p transformation, it follows that the ESP expected for the LT will be satisfied. In other words, it is predicted that the observer will always find that the energy and inertial mass of objects moving with respect to him have values that are γ times greater than the corresponding values obtained when they are stationary in his rest frame. It would therefore appear that the subjective character of STR does exist for these properties, even though it has been shown to be violated (see Chapter 4.A) for space-time measurements. *This conclusion overlooks an important assumption made in the derivation of eq. (85),* however, as will be discussed in the following section.

C. Uniform Scaling of Physical Properties

The derivation of the space-time transformation assumes that the rest frames S and S′ are inertial and are therefore moving with constant velocity relative to each other. By contrast, the derivations of the energy-momentum relations given above assume that a constant force has been applied to an object at some definite origin, causing it to be accelerated in accordance with Newton's Second Law. In particular, the relations of eqs. (78-79) for inertial mass and energy are obtained for the special case when the observer in S′ is co-moving with the object ($\mathbf{dp'} = 0$) and his counterpart in S is stationary at the location where the force was applied. *Under the circumstances it is possible to distinguish S′ from S, and therefore one cannot claim that Einstein's ESP applies in this case.*

The fact is that there are grounds for criticizing both Einstein's original derivation[5] of the relativistic E-p relations in Chapter 7. A and the alternative approach based on the transformation of eqs. (85-86) in Chapter 7. B. In the last analysis such a theoretical analysis is always subject to experimental scrutiny. The dependence of inertial mass on its speed relative to the observer given in eq. (78) has been convincingly verified by experiments Bucherer[66] carried out shortly after Einstein's paper[5] appeared. It needs to be clearly understood that these investigations only confirm that the inertial mass m of a particle increases in accord with eq. (78) when it is accelerated to speed v relative to the observer in the laboratory. There is no comparable evidence to confirm the ESP in this case, *i.e.* to demonstrate that the inertial masses of objects that are stationary in the laboratory rest frame S are measured to be larger than their rest mass values by the same factor γ from the vantage point of the rest frame S′ of the accelerated particles.

The experience with atomic clocks needs to be considered in this connection. All experimental data to date are consistent with the UTDL of eq. (57) which stands in contradiction to the ESP. The experiments that have been done[66] to study the variations in inertial mass indicate that it varies at exactly the same rate as the periods of atomic clocks. A reasonable conclusion is therefore that *the same proportionality relationship holds for both quantities* under all conditions. On this basis, it is reasonable to expect that inertial masses satisfy the following relation:

$$m_2\, \gamma(v_{10}) = m_1\, \gamma(v_{20}). \tag{93}$$

Note that the order of indices is not the same as in eq. (57), because the τ_i quantities refer to elapsed times there rather than the corresponding periods of the clocks used to measure them. The above discussion emphasizes that the fractional change in the lifetimes of particles and other periods of proper clocks is exactly the same as for their inertial masses. The simple proportionality in eq. (59) between measured times in S′ and S has advantages in this respect. Thus, the corresponding relation between measured inertial masses is:

$$m' = mQ^{-1}, \tag{94}$$

whereby Q has the same value as for clock periods and lifetimes. If the clock in S′ runs slower, then $Q > 1$; the opposite relation holds when the S′ clock runs faster ($Q < 1$).

The simplest way to avoid confusion in this respect is to think in terms of units. The units of time, length/distance and inertial mass increase with the speed of the corresponding rest frame relative to the appropriate reference system (ORS; see Chapter 5.C). The parameter Q is a conversion factor between the units employed in S and S′. Accordingly, $Q > 1$ if the speed of S′ relative to the ORS is greater than that for S, and $Q < 1$ if the opposite relationship holds. The measured values expressed in these units are therefore smaller in each case than the corresponding values obtained by a stationary observer in the ORS.

The analogous relation for distances l and l′ is accordingly:

$$l' = lQ^{-1}. \qquad (95)$$

Measurement is objective in the theory, so the units of other quantities can be obtained by appropriate algebraic manipulation based on their composition in terms of the fundamental properties of inertial mass, length and time. For example, since speed/velocity v is a ratio of a distance to a corresponding elapsed time, it follows that

$$\mathbf{v}' = Q^0\mathbf{v} = \mathbf{v}, \qquad (96)$$

which is of course consistent with the light-speed postulate. Linear momentum **p** is equal to the product of inertial mass and velocity, hence

$$\mathbf{p}' = \mathbf{p}Q^{-1}. \qquad (97)$$

Angular momentum $\mathbf{p}_\Theta$ is defined as the product of a distance with a momentum value, and thus the corresponding relation in this case is:

$$\mathbf{p}_\Theta' = \mathbf{p}_\Theta Q^{-2}, \qquad (98)$$

i.e. the unit of angular momentum varies as Q^2. The Ives-Stilwell experiment[34] provides a key test of this scaling.[54] Because of the $E = mc^2$ relation of eq. (80), it follows from eq. (96) that energy E scales in the same manner as m:

$$E' = EQ^{-1}. \qquad (99)$$

At the same time, the above experiment shows that the frequency ν of the moving light source is γ times smaller than the corresponding *in situ* value. This result is naturally consistent with the UTDL and the scaling of time in eq. (59), with $Q = \gamma$ in this case. As a consequence, the following general result holds for the scaling of frequency ν since it is the reciprocal of the period of the radiation:

$$\nu' = \nu\, Q. \qquad (100)$$

The ratio of the energy of the light waves to the corresponding frequency is given by Planck's Law:

$$E = h\nu, \tag{101}$$

where h is Planck's constant. Experimental tests of eq. (101) are normally carried out *in situ, i.e.,* with the light source stationary in the rest frame of the observer. The Ives-Stilwell experiment raises the question of whether this law also holds when the light source is moving relative to the observer. The above equations are in fact consistent with this supposition, even though energy and frequency do not scale in the same manner. This is because the constant h in eq. (101) has the units of angular momentum and therefore must scale as Q^2 according to eq. (98). This means that in the Ives-Stilwell experiment *the value of Planck's constant for the observer in the laboratory S is actually $\gamma^2 h$ in the rest frame S′* of the moving light source. The observer at rest in S′ measures the standard values of E′ and ν′ for the light source, and thus finds that $E' = h\nu'$ since Planck's constant has the normal value under these conditions. According to eqs. (99-100), his counterpart at rest in the laboratory must find that

$$E = \gamma E' = \gamma^2 h\,(\gamma^{-1}\,\nu') = \gamma h\nu', \tag{102}$$

which is therefore also in agreement with Planck's Law. Unless one assumes that the constant h scales in the above manner, the conclusion would have to be that Planck's Law is only valid for *in situ* measurements, which would be an unnecessary restriction on its theoretical significance.

The discussion of acceleration **a** in STR is rather complicated and usually proceeds by introducing a so-called "four-velocity" defined as $\mathbf{w} = \gamma\mathbf{v}$.[67] It leads to the following relations between the components of acceleration **a′** in the instantaneous rest frame of the particle (S″) and their values measured in another rest frame:

$$\begin{aligned} a_x &= \gamma^{-3} a_x'' \\ a_y &= \gamma^{-2} a_y'' \\ a_z &= \gamma^{-2} a_z''. \end{aligned} \tag{103}$$

The transformation is only valid for the instantaneous rest frame, which by definition is not inertial, so it is not relativistic in the strict sense. The equations are reminiscent of Einstein's distinction between longitudinal mass and transverse mass mentioned in Chapter 7. A.

Table 1. Variation of the units of physical quantities with state of motion. The ratio of the respective units of a given quantity in different inertial systems is conveniently given as a power n of the parameter Q in the ALT of eq. (62).

Physical Quantity	Standard Unit (mks system)	Power of Q n
Time	s	1
Length	m	1
Energy	J = Nm	1
Force	N	0
Velocity (speed of light)	m/s	0
Gravitational Mass	kg	0
Inertial Mass	Ns^2/m	1
Momentum	Ns	1
Acceleration	m/s^2	−1
Angular Momentum (Planck's constant)	Nms	2
Torque	Nm	1
Radiative Frequency	s^{-1}	−1

The alternative scaling relationship suggested by the above discussion is obtained quite simply from eqs. (59, 96) and the definition of $\mathbf{a} = d\mathbf{v}/dt$:

$$\mathbf{a}' = Q\,\mathbf{a}. \tag{104}$$

There is no distinction in the scaling for different components and no restriction as to the identity of either rest frame. The corresponding scaling for force **F** is obtained uniquely from Newton's Second Law as either d**p**/dt or m**a**, namely as

$$\mathbf{F}' = Q^0\mathbf{F} = \mathbf{F}. \tag{105}$$

Just as for velocities in eq. (96), the values of force components are seen to be independent of the scaling factor Q and are thus the same for all observers. It needs to be emphasized that the above scaling relationships are obtained under the assumption that all observers and objects of their measurements are at the same gravitational potential. This restriction will be removed in Chapter 9.

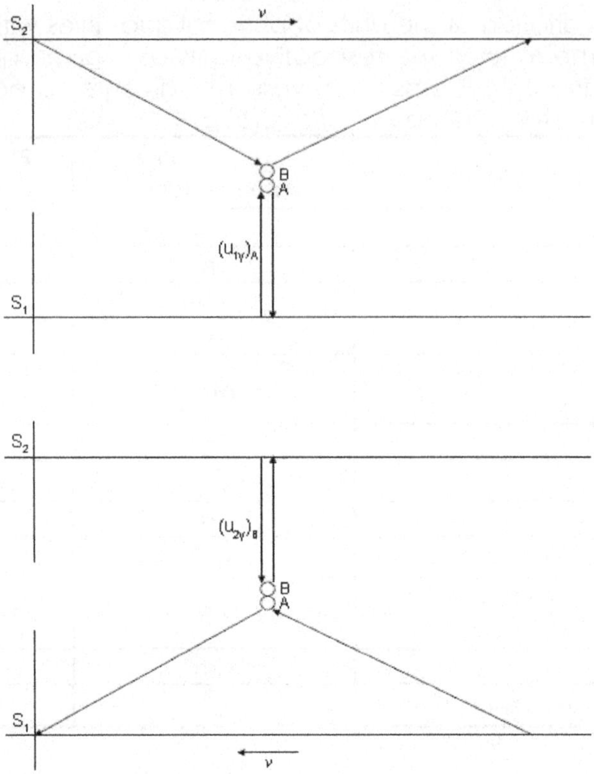

Fig. 3. Lewis-Tolman model for the elastic collision of two objects that were originally at rest in different inertial systems S_1 and S_2 that are moving with speed v relative to one another. The upper half of the diagram shows the collision as viewed by an observer at rest in S_1 (note that S_2 is moving to the right for him). The lower half shows the same collision as viewed by an observer at rest in S_2 (he views S_1 moving to the left).

In summary, it has been argued in the above discussion that all dynamical properties are subject to a *uniform scaling* that relates the corresponding measured values obtained by observers in different rest frames. An integral power of the parameter Q in the ALT of eq. (62) serves as a conversion factor in each case, as listed in Table 1. The latter can be obtained from the UTDL of eq. (57) by comparing the rates of proper clocks which are stationary in the rest frames being compared. The speeds in the corresponding γ factors are defined relative to a specific rest frame (ORS; Chapter 5. C)[47] in each case. An objective theory of measurement is assumed for the above conclusions. It is consistent with the

Principle of Rational Measurement (PRM; Chapter 4. A), which like the UTDL, can be taken as an additional postulate for the present version of relativity theory.

D. Conservation of Momentum and the Lewis-Tolman Conjecture

One of the interesting questions that arose because of Einstein's theory[5] was how it affects the conservation laws of classical physics. It is clear from eq. (78), for example, that the law of conservation of mass is not valid for high-energy collisions. The γ factors that multiply the rest masses in such equations are potentially different for each collision partner before and after the interaction and this circumstance clearly prevents satisfaction of this law on a sufficiently general basis.

It is important to note that, although eq. (78) can be deduced from eq. (77), it was not stated explicitly in Einstein's 1905 paper. Credit for this theoretical result is due to Lewis and Tolman,[68] and their work provided the impetus for the experiments which confirmed mass dilation. They presented a Gedanken experiment which is based on both time dilation and the FLC, as discussed below. The basic content of their theoretical arguments can be understood with the help of the diagram in Fig. 3. An elastic collision between two identical objects A and B is considered from the vantage point of two observers who are in relative motion with speed v. The observers are located in inertial systems S_1 and S_2, respectively, and the direction of their relative motion is along the x-axis. Object A is initially at rest in S_1 before the collision, whereas object B is initially at rest in S_2. The collision process is designed to be perfectly symmetrical for the two observers. Each one sees one of the objects moving along the y axis, making a collision with the other object after it travels a distance y relative its initial rest position. It then returns to its initial position with exactly the same speed. The other object appears to be traveling at (high) speed v along the x (horizontal) direction but with a small (vertical) component so that it makes a glancing collision with the former object. After the collision, the other object continues moving along the x-axis in the same direction but returns with a vertical component in the opposite direction. The underlying idea is that the latter object will simply appear to move up and down in the y direction for the other observer.

The top part of Fig. 3 shows how the collision plays out for the observer in S_1. For him object A moves upward with speed u_{1y}, makes the collision with B, and then returns with the same speed. The time for the object to move before the collision occurs is T_0, and hence $(u_{1y})_A = y/T_0$. The bottom part shows the collision from the perspective of the observer in S_2. He finds that object B moves downward with speed $(u_{2y})_B = y/T_0$ before returning to him in the opposite direction. The clear assumption is therefore that $(u_{1y})_A = (u_{2y})_B$ in Fig. 3.

Let us now consider how the motion of object B appears to the observer in S_1. It is here that Lewis and Tolman begin to make use of some of Einstein's conclusions in his original work.[5] First, they assumed that because of time dilation in S_2, the time T of the downward flight of object B must be longer than T_0 by a factor of $\gamma = (1 - v^2/c^2)^{-0.5}$. Secondly, because of the FLC, the distance traveled by the object over this period is equal to y from the observer's vantage point in S_1, that is, the same distance as measured by the observer in S_2. This is because the direction of the object's motion is *perpendicular* to that of the relative motion of the two inertial systems, in accord with the FLC in eq. (21). As a result, Lewis and Tolman concluded that the speed of object B is γ times less than that of object A, again from the viewpoint of the observer in S_1, that is, $(u_{1y})_B = y/T = \gamma^{-1}y/T_0 = \gamma^{-1}(u_{1y})_A$.

It is at this point in the discussion that conservation of momentum is brought in. If one assumes that the respective y components of the momenta of the two objects must be equal in order for object A to return to its starting position with the same speed as before $[(u_{1y})_A]$, it follows that the inertial mass of object B (as measured in S_1) must be *larger* than that of object A by the same factor of γ, in agreement with eq. (78). In short, Lewis and Tolman concluded that the two objects no longer have the same inertial mass by virtue of the fact that S_2 is moving with speed v relative to the observer in S_1.

The key point in the above arguments is that the two observers disagree on the values of the speeds traveled by the two objects, in contradiction to eq. (96). The observer in S_1 finds that object B has a downward component of velocity $[(u_{1y})_B]$ of magnitude y/T, whereas his counterpart in S_2 measures a value of y/T_0

for the same quantity $[(u_{2y})_B]$. Despite the fact that this assumption leads to the correct prediction of mass dilation, there are nonetheless clear problems with it. There is no restriction in the above discussion regarding the nature of the two objects, and so one critical possibility is to use single photons for this purpose. In this case there is no question about the speed of object B for either observer. It must be equal to c for both of them, and hence the Lewis-Tolman condition of $(u_{2y})_B = \gamma(u_{1y})_B$ in Fig. 3 cannot hold in this case.

Moreover, the above condition is violated on a much more general basis in subsequent discussions that have appeared in the literature regarding the effect of time dilation for meta-stable particles moving at relativistic speed.[43,44] In this example it is universally assumed that the speed measured for these particles is *independent* of the state of motion of the observer, in agreement with eq. (96). The above observation shows at the very least that there is a significant inconsistency in the way relativistic arguments are applied in these two examples.

There is therefore a need to consider what consequences result from removing this assumption from the Lewis-Tolman Gedanken experiment for mass dilation. First of all, there is no empirical evidence to suggest that the type of collision experiment described in Fig. 3 actually plays out in the way Lewis and Tolman have claimed. If it is assumed instead that $(u_{2y})_B = (u_{1y})_B$, it is clear that the only way to satisfy conservation of momentum in this case is to either reduce the rest mass of object B by a factor of γ or to lower its downward speed by the same proportion. If B is a light pulse, the latter option is clearly unavailable, but the desired decrease in momentum can still be achieved in this case by reducing its energy by the same factor. The conclusion is that in order to satisfy the principle of momentum conservation in the Lewis-Tolman model, it is necessary to change some characteristic of one of the collision partners, either its velocity or its inertial mass. Their conclusion on the basis of the above model that inertial mass satisfies eq. (78) is therefore fortuitous, but it is nonetheless correct based on the results of subsequent experiments.

This experience is a reminder that experiment never *proves* the validity of a theoretical argument used to successively predict its results. It can be used to disprove it, however, if the results are shown to be inconsistent with the theoretical prediction.

Newton's Second Law also cannot be proven, nor the Law of Energy Conservation. All indications from empirical evidence are that they are both correct, however. The principle of momentum conservation follows from the Second Law and its conclusion that $d\mathbf{p}/dt = 0$ if no external force $\mathbf{F}$ is applied. Thus the sum of the momenta of the particles before (p_{ib}) and after (p_{ia}) their interaction must be the same:

$$\Sigma\, \mathbf{p}_{ib} = \Sigma\, m_{ib}\mathbf{u}_{ib} = \Sigma\, m_{ia}\mathbf{u}_{ia} = \Sigma\, \mathbf{p}_{ia}. \tag{106}$$

The above equation nonetheless emphasizes that a feature of the non-relativistic treatment of momentum conservation must be abandoned for high-energy collisions. Since the conservation law must hold for all observers, it is argued that the momenta can be computed relative to an *arbitrary rest frame* moving with speed v relative to the location of the collision:

$$\Sigma m_{ib}(\mathbf{u}_{ib} - \mathbf{v}) = \Sigma m_{ia}(\mathbf{u}_{ia} - \mathbf{v}). \tag{107}$$

The above equation is only consistent with eq. (106) if the sum of the inertial masses is also conserved, however, and therefore is invalid in the high-energy regime.

Belief that the principle of conservation of momentum is nonetheless valid for all observers independent of their state of motion is nonetheless supported by the concept of uniform scaling discussed in Chapter 7. C. The above argument makes clear, however, that the particle velocities in eq. (106) must be defined relative to a *unique* reference frame, namely that in which the location of the collision is fixed in space. In that case, one can apply the scaling relationships for inertial mass and velocity in eqs. (94,96) to guarantee that eq. (106) will hold for an observer who is moving with respect to that origin. In effect, that observer simply uses a different set of rational units to compute the values of the momenta for each particle before and after the collision. This restriction in the way the velocities of the particles must be defined in eq. (106) is consistent with the method used to deduce the scaling relationships. This is because the velocity $\mathbf{v}$ used in these derivations [see eq. (83)] is always taken relative to the origin where the force $\mathbf{F}$ is applied, in accord with both Hamilton's canonical equations and Newton's Second Law.

The same argument also holds for conservation of energy in collisions and elsewhere. Two observers in relative motion will not agree on the value of a given energy because of eq. (79), but this is only because they use a different standard unit in their computations. If the sums of the particle energies before and after the collision are equal for one observer, they will also be equal for any other regardless of their relative state of motion. The above relationships are clearly consistent with the amended RP given in Chapter 6. C. The different observers are not aware of any distinctions in the units they employ based on their respective *in situ* observations, but it becomes evident that this is the case once they compare their measured values for the same object. The scaling relations given in Table 1 allow them to quantitatively predict the ratios of their respective measured values for any quantity once they have determined the value of the scaling parameter/conversion factor Q in the ALT.

Chapter 8

Relativistic Theory of Electromagnetism

Distances can be measured using atomic clocks by making use of Einstein's light-speed postulate, as discussed in Chapter 4. B. This procedure finds practical application in the Global Positioning System (GPS). Clocks on the satellites are slowed by time dilation (after correcting for the effects of the gravitational red shift[59]), but consistent with the RP no change in rate will be noted onboard. The same holds true for length measurements, but the elapsed times for light to traverse the objects on the satellite will be greater on clocks on the earth's surface (again neglecting gravitational effects) than they were prior to launch. Since the speed of light does not change on the satellite, in accord with Einstein's postulate[5] of the constancy of the speed of light in free space, it follows that the objects' dimensions have *expanded* isotropically as a result of the acceleration [see eq. (58) and the discussion in Chapter 5. D]. As shown in Chapter 4. B, the latter result runs contrary to the predictions of the FLC and therefore reveals a lack of internal consistency in STR (clock riddle [32]).

The problem is corrected by introducing a different space-time transformation than the LT, one which assumes the proportionality of clock rates used in the GPS methodology (ALT; see Chapter 6. A). However, one of the main purposes of STR was to construct a relativistic theory that is compatible with the Maxwell electromagnetic field equations. Einstein's two postulates were chosen specifically for achieving this goal. The question is thus how making what amounts to an additional postulate in the revised theory affects the treatment of electromagnetic interactions.

In particular, it has been shown in Chapter 7. C that assuming strict proportionality between the rates of clocks in relative motion implies that all other physical properties are subject to similar relationships. Since time, energy and distance have equal proportionality constants for observers in different rest frames [see eqs. (59,95,99) and Table 1], it follows that there must always be complete agreement on their respective values for forces and relative velocities [eqs. (96,105)]. This conclusion again runs contrary to STR since it holds that the values of forces in directions that are transverse to the direction of the relative velocity of the two rest frames will be different for the respective observers, although the corresponding parallel components will be the same.[5,69]

A review of the relativistic theory of electromagnetic interactions will be undertaken in the following discussion. Emphasis will be placed on the way the revised mechanical theory based on the GPS methodology differs from conventional STR, and the implications this has on the treatment of Maxwell's electromagnetic field equations.

A. Critique of Einstein's Derivation of the Equations of Motion

The starting point in the discussion is the Maxwell field equations in free space:

$$c^{-1}\partial E_x/\partial t = \partial B_z/\partial y - \partial B_y/\partial z - c^{-1}\partial B_x/\partial t = \partial E_z/\partial y - \partial E_y/\partial z$$

$$c^{-1}\partial E_y/\partial t = \partial B_x/\partial z - \partial B_z/\partial x - c^{-1}\partial B_y/\partial t = \partial E_x/\partial z - \partial E_z/\partial x \quad (108)$$

$$c^{-1}\partial E_z/\partial t = \partial B_y/\partial x - \partial B_x/\partial y - c^{-1}\partial B_z/\partial t = \partial E_y/\partial x - \partial E_x/\partial y,$$

where E_i and B_i ($i = x, y, z$) are the respective electric and magnetic field components, x, y, z and t are the spatial and temporal coordinates and c is the speed of light in free space. These equations are valid in all inertial systems, and in accord with the RP they must have the same form in each. On this basis Einstein assumed that the above equations must be invariant to a suitable space-time coordinate transformation. This is a key assumption that turns out to be unnecessarily restrictive, as will be discussed in Chapter 8. D. Maxwell also showed that the speed of light must be the same in all inertial systems in order for the equations to have general validity, thus ruling out the classical GT for this purpose. The LT does satisfy this criterion but without having additional information other than the constancy of the light speed c, it is only possible to specify the exact form of the corresponding equations to within a normalization factor ε in the GLT of eq. (15). As pointed out in Chapter 2. B, since speed is a ratio of the change in distance travelled to the corresponding elapsed time, the same result for the light speed holds for any value of ε. Einstein was aware of this lack of specificity,[5] but he argued that ε can only depend on the relative speed v of the two inertial systems S and S′ related in the transformation, and on this basis concluded that $\varepsilon = 1$ is the only acceptable value for this parameter. On the contrary, it has been shown in Chapter 6. A that the GPS assumption of the proportionality of the rates of clocks in relative motion can be incorporated into the GLT by setting $\Delta t'$ in eq. (59) to $Q^{-1}t$, where Q is the value of the proportionality constant used in the GPS "pre-correction"[59] procedure. On this basis the value of ε is found to be $\eta(Q\gamma)^{-1} = (1 - v\Delta x c^{-2}\Delta t)^{-1}(Q\gamma)^{-1} = (1 - vu_x c^{-2})^{-1}(Q\gamma)^{-1}$, which depends on both the velocity component $u_x = \Delta x/\Delta t$ of the object of the measurement and the relative speed v of S and S′. The result is the ALT of eq. (62).

In order to apply any version of the GLT to the Maxwell equations given in eq. (108), it is first necessary to obtain the corresponding relations between partial derivatives[70] of the four coordinates [for simplicity, the transformation is expressed in terms of the space-time coordinates x,y,z,t rather than the intervals used in eq. (15)]:

$$\begin{aligned} \partial/\partial x &= \gamma\varepsilon'(\partial/\partial x' - vc^{-2}\partial/\partial t') \\ \partial/\partial y &= \varepsilon'\partial/\partial y' \\ \partial/\partial z &= \varepsilon'\partial/\partial z' \\ \partial/\partial t &= \gamma\varepsilon'(-v\partial/\partial x' + \partial/\partial t'). \end{aligned} \tag{109}$$

Because of the homogeneous form of the Maxwell equations in eq. (108), it is clear that the value of the function ε in the above equations is immaterial for the outcome of applying this transformation to obtain the corresponding relations in S′. This fact demonstrates that the GPS-compatible form of the GLT eq. (62), with $\varepsilon' = \eta'(Q'\gamma)^{-1}$, is just as suitable for carrying out the transformation between different rest frames as Einstein's LT.[5] It is therefore not possible to distinguish between the two versions of relativity theory on this basis.

The next step in Einstein's derivation is to find a set of relations coupling the respective electromagnetic field components in the two rest frames. He assumed that the relationship must be such that when the corresponding substitutions are made in the transformed version of eq. (108), the result is an equivalent set of Maxwell field equations in S′. Einstein succeeded[5] in finding the desired relationships, but as before with the derivation of the LT, it is found that they also come with an undetermined normalization constant, which he referred to as ψ(v):

$$\begin{aligned} &E_x'=\psi(v)E_x && B_x'=\psi(v)B_x \\ &E_y'=\gamma\psi(v)(E_y-vc^{-1}B_z) && B_y'=\gamma\psi(v)(B_y+vc^{-1}E_z) \\ &E_z'=\gamma\psi(v)(E_z+vc^{-1}B_y) && B_z'=\gamma\psi(v)(B_z-vc^{-1}E_y). \end{aligned} \tag{110}$$

In this case he argued that ψ(v) must be an even function of v and *also be the reciprocal of ψ(–v)*. The conclusion on this basis is that the only permissible value is ψ(v) = 1.

In order to deduce the transformation properties of the law of motion for the electromagnetic interaction it is necessary to also specify how the components of acceleration transform. Einstein used the transformation in eq. (103) for this purpose (with S″ = S†). He thus obtained the following equations of motion for the stationary observer (μ is the rest mass of the electron and e is its electric charge) by combining eqs. (110) and (103):

$$\begin{aligned} &F_x\dagger = \mu a_x\dagger = \mu\gamma^3 a_x = \mu\gamma^3 d^2x/dt^2 = eE_x\dagger = eE_x \\ &F_y\dagger = \mu a_y\dagger = \mu\gamma^2 a_y = \mu\gamma^2 d^2y/dt^2 = eE_y\dagger = e\gamma(E_y - vc^{-1}B_z) \\ &F_z\dagger = \mu a_z\dagger = \mu\gamma^2 a_z = \mu\gamma^2 d^2z/dt^2 = eE_z\dagger = e\gamma(E_z + vc^{-1}B_y). \end{aligned} \tag{111}$$

He assumed thereby that the equations of motion in the rest frame (S†) *in which the electron is stationary* can be obtained on the basis of the non-relativistic (**F** = m**a**) version of Newton's Second Law. He also assumed that that the electronic charge e is invariant and that only electric fields need to be considered in S† since the electron is stationary there.

The above derivation of eq. (111) raises a number of questions, however. The most important one is whether it is permissible to apply the RP in the present case *given that S† is not an iner-*

tial system. The electron is under constant acceleration and therefore is not in a state of uniform translation. The assumption that the observer is moving along with the electron does not change this basic point. It simply shows that the observer at rest in S† is also not in uniform translation.

The derivation of eq. (103) that Einstein used for the transformation of the acceleration components also merits closer consideration. An alternative means of obtaining these relationships would seemingly start with the relativistic velocity transformation (VT) of eq. (29). The latter is obtained by division of the variables in the GLT of eq. (15), and therefore also those in the LT and the ALT. In the case under discussion, $u_x = x/t = v$, and thus $\eta = (1 - vx/c^2t)^{-1} = \gamma^2$. However, **u†** *has a constant null value* by virtue of the definition of S† as the rest frame in which the electron is always at rest. This means that all three **a†** components would also have null values for the stationary observer in S†. When looked upon in this light, the left-hand sides of all three relations in eq. (105) are therefore also zero, thus invalidating Einstein's derivation entirely.

Perhaps in the spirit of the above remarks, Planck suggested a different approach, as has already been discussed in Chapter 7. A, by noting[38] that the electromagnetic force components F_i can be obtained by using Newton's Second Law ($\mathbf{F} = d\mathbf{p}/dt$) in conjunction with the relativistic definition of momentum given in eq. (81),

$$F_x = d/dt\,(\gamma\mu v_x) = \mu\gamma^3 d^2x/dt^2 = \mu\gamma^3 a_x$$
$$F_y = d/dt\,(\gamma\mu v_y) = \mu\gamma d^2y/dt^2 = \mu\gamma a_y \qquad (112)$$
$$F_z = d/dt\,(\gamma\mu v_z) = \mu\gamma d^2z/dt^2 = \mu\gamma a_z.$$

The equations of motion in Einstein's eq. (111) then result by introducing the expression for the Lorentz force,[71] $\mathbf{F} = e(\mathbf{E} + \mathbf{v}c^{-1}\mathbf{x}\mathbf{B})$, with **v** parallel to the x-axis:

$$F_x = \mu\gamma^3 a_x = \mu\gamma^3 d^2x/dt^2 = e\,E_x\,,$$
$$F_y = \mu\gamma a_y = \mu\gamma d^2y/dt^2 = e\,(E_y - vc^{-1}B_z) \qquad (113)$$
$$F_z = \mu\gamma a_z = \mu\gamma d^2z/dt^2 = e\,(E_z + vc^{-1}B_y).$$

It is important to note that neither eq. (103) nor (110) is required in the derivation of eq. (113). On this basis there is no justification for making a distinction between "longitudinal mass" ($\gamma^3\mu$) and "transverse mass" ($\gamma^2\mu$) in the equations of motion. The factors multiplying the rest mass evolve in a straightforward way as a result of differentiating the relativistic momentum to obtain eq. (113).

B. Relativistic Transformation of Forces

Although the equations of motion in eqs. (111) and (113) are identical, there is nonetheless an important distinction in their corresponding transverse relations. There is an additional factor of γ on the right-hand sides for the y and z components in eq. (111). The latter factor is cancelled by the same factor for the corresponding second-derivatives with respect to the time coordinate on the left-hand sides. The reason for this distinction can be seen by comparing the left-hand sides of the two sets of relations. In eq. (111) Einstein used the force components $F_i†$ for the stationary observer in S†, whereas eq. (113) has the corresponding force components F_i for the observer who is at rest in S. Thus, comparison of these two sets of equations indicates that there is an asymmetry between the force components in the two rest frames:[69]

$$F_x† = e\,E_x† = F_x$$
$$F_y† = e\,E_y† = \gamma F_y \qquad (114)$$
$$F_z† = e\,E_z† = \gamma F_z.$$

This theoretical result is especially interesting in light of eq. (105) for the scaling of force components. The two sets of components are expected to be the same for observers in all inertial systems on this basis. Since forces are ratios of energy and distance, it follows from eqs. (95) and (99) that their measured values are the same for observers in different rest frames, in agreement with eq. (105).

In order to resolve the discrepancy between eq. (114) and eq. (105), it is important to go back to Einstein's derivation of the equations of motion in eq. (111). It is not true that S†, which is the rest frame of the accelerated electron, is an inertial system. As already mentioned in the previous section, this circumstance calls into question the applicability of the RP in this case. The rest frame in which the electromotive force is applied to the electron is distinguishable from all other inertial systems. The situation is the same as in Einstein's discussion of an electron moving in a closed path and returning to the origin with less time having elapsed on its co-moving proper clock.[5,14] The extra factor of γ in eq. (111), as well as the asymmetry in eq. (114), therefore depends on a faulty assumption.

Beyond this point, there is a more general question regarding the implications of the RP in assessing the validity of physical laws. Since Einstein's work,[5] it has been assumed that such laws must be invariant to an LT to another rest frame, and that the simplest way to guarantee this requirement is that the laws have a *covariant* form. There is another obvious way to achieve the same end, however, which is underscored by the amended version of the RP discussed in Chapter 6. C. Physical laws are equations that must conform to a consistent choice of rational units to express them. When there is a unique rest frame where interaction A occurs, as is the case in the present discussion of electromagnetic interactions, it is sufficient as far as the RP is concerned if the same law holds in every other inertial rest frame when the *analogous (and therefore different) experiment B* is carried out there. Moreover, if an observer in a different inertial system carries out measurements for interaction A, it is sufficient for the RP if his results differ from those of the local observer *by only a change in their respective units*. In the present case this means that the velocity **v** of the charged particle used to obtain the Lorentz force and also its momentum from eq. (81) must be determined relative to the unique rest frame (the ORS[47] in this example; see Chapter 5. C) in which the electromotive force is applied. The speed of the electron relative to an arbitrary rest frame in which a given observer is stationary plays no direct role in this determination. Because of eq. (96), the value of **v** will be the same for all observers, regardless of their state of motion, but the measured results for other quantities will differ in accord with the other scaling relationships shown in Table 1. The scaling parameter in each case is a power of the conversion factor Q in the ALT connecting the observer's rest frame with any other where measurements of the same quantities have been made.

The above interpretation removes any questions in Einstein's original derivation[5] of the equations of motion for the electron that arise because the S† rest frame is not inertial. It also eliminates the need to rely on eq. (103) to determine the relationships between the acceleration components in S and S†, thereby invalidating the asymmetric relation between the respective components of the electromagnetic force in eq. (114). Instead one simply uses Newton's Second Law to obtain the relations between the Lorentz force components and the corresponding acceleration values needed to deduce the equations of motion of eq. (113) in the ORS. The scaling relationships of Table 1 can then be used to determine the measured values of other physical quantities in any other rest frame based on the ORS results. In particular, it is assumed on the basis of eq. (105) that each observer obtains the same values for the force components as in the latter rest frame.

C. Test Example Employing the Uniform Scaling Relationships

The theoretical treatment discussed in the preceding section makes a definite assumption about the relative rates of clocks in S and S†, specifically that the one in S† moving with the electron runs γ times slower than its counterpart in S. An important aspect of the ALT version of relativity theory is that the ratio of clock rates between arbitrary rest frames can take on any value, consistent with the proportionality axiom of eq. (59). Thus, by virtue of the UTDL in eq. (57), $Q = \gamma > 1$ relative to the ORS. Since the revised theory is objective and conforms to the PRM (see Chapter 4. A), it is necessary to assume that the inverse relationship holds from the standpoint of the observer in S, namely

$$\Delta t = Q\, \Delta t' = Q'^{-1} \Delta t' \tag{115}$$

A simple way to think about these relationships, as discussed in more detail in Chapter 7. C, is to ascribe a different unit of time in the two rest frames. Because of the RP, both observers assume that their local unit is standard (1 s), but in reality only one (or neither) of them can make this claim. In other words, the standard has to be defined in a specific rest frame which is mutually agreed upon to be fully consistent with observation. In the case of the GPS methodology it is natural to define the standard unit of 1 s to a clock at a particular location on the earth's surface. The corresponding unit on the satellite (excluding the effects of the gravitational red shift) based on this standard would then be Q s. This means that the numerical value of an elapsed time Δt for an event anywhere in the universe will be Q times greater at the standard location than that measured locally on the satellite ($\Delta t'$), consistent with eq. (115). In other words, Q can be looked upon as a *conversion factor* that must be used in order to predict elapsed times on the ground based on measurements carried out with an identical (proper) clock located on the satellite (S′). Analogous considerations hold for all other physical properties [see eqs. (94-102) and Table 1.

A good example for illustrating the above points can be given with reference to the force equations in eq. (112). It is important to recognize that μ is the rest mass of the electron in S′ and thus not the inertial mass m of the electron when observed in a rest frame *other than S′*. The general formula for this quantity depends on the value of the conversion factor in the rest frame S (not necessarily the ORS) in which the measurements of the event in S′ are to be made. It helps to avoid confusion by rewriting the equations in terms of the variable $m = Q\mu$; for example, $F_y = ma_y = Q\mu a_y$. It is then understood that both m and a_y are to be determined by an observer in his units. If the observer is at rest in S′, then $m = \mu$, since $Q = 1$ there.

The reason why it is important to make the above distinction is that it has an important effect on the equation of motion for F_x due to the γ^3 factor in eq. (112). The latter factor arises because of the differentiation of the momentum using Newton's Second Law. In particular, a factor γ^2 comes from $d\gamma/dt$, whereas the remaining factor of γ comes from the value of the inertial mass of the electron in the observer's units. Thus, in light of the above remarks, it is preferable to rewrite the equation in terms of m, namely as:

$$F_x = \gamma^2 m a_x. \qquad (116)$$

This conclusion is of special interest when evaluating eq. (115) in S′, *i.e.* the rest frame of the electron: $F_x = \gamma^2 \mu a_x' = \gamma^2 \mu d^2x'/dt'^2$. In deriving eq. (103), Einstein assumed to the contrary that $F_x = \mu a_x'$, but in doing so, he overlooked an important aspect of the problem at hand. By construction, the electron is under constant acceleration along the x-axis, and thus its momentum is changing with time. Einstein's formula is only valid for the case in which the speed v_x is constant *until the force is applied* to produce the acceleration.

To illustrate how these conclusions work out in practice, let us next consider the case of two electrons moving in parallel along the x-axis. For an observer in an arbitrary rest frame S, the equation of motion using the Lorentz force in the y-direction is:

$$F_y = dp_y/dt = ma_y = Q\mu d^2y/dt^2. \quad (117)$$

The formula for the time-dependence of the change Δy in the location of one of the electrons is therefore:

$$\Delta y = F_y \, \Delta t^2/2Q\mu. \quad (118)$$

The scaling of distance in eq. (95) requires that the value in S′ in which the electrons are at rest must be $\Delta y' = Q^{-1}\Delta y$. The corresponding equation of motion is [with $\Delta t' = Q^{-1}\Delta t$ from eq. (115)]:

$$F_y' = \mu a_y' = \mu d^2y'/dt'^2 = F_y, \quad (119)$$

from which it follows from eqs. (117-118) that

$$\Delta y' = F_y \, \Delta t'^2/2\mu = F_y \, Q^{-2}\Delta t^2/2\mu = Q^{-1} \, \Delta y, \quad (120)$$

in agreement with expectations based on eq. (95). Thus the distance values measured locally there are always smaller than for a stationary observer in the laboratory (ORS with $Q = \gamma > 1$) from which the electron was accelerated. This is because the unit of distance (meter stick) is greater in S′, not because there is length contraction in that rest frame. The same general result in eq. (120) holds for a stationary observer in any rest frame, *i.e.*, with $Q \neq \gamma$. Note that Q may be less than unity, in which case the stationary observer in S will obtain smaller distance values than his counterpart in S′ moving with the electron.

Next consider the equation of motion in the parallel (x) direction:

$$F_x = dp_x/dt = m\gamma^2 a_x = Q\gamma^2\mu d^2x/dt^2. \quad (121)$$

Note the distinction between Q and γ in this equation. It is necessary to have three distinct rest frames to describe the most general case: the ORS where the initial acceleration occurs, the (non-inertial) rest frame S′ co-moving with the electron, and an arbitrary rest frame S in which the observer is stationary. The value of $\gamma(v)$ is determined from the speed v of the electron relative to the ORS. By contrast, the observer is assumed to be stationary in a rest frame S where proper clocks run Q times faster (if $Q > 1$) or slower (if $Q < 1$) than those at rest in S′. Note that the conversion factor between clock rates in the ORS and S does not occur in the present calculation. The formula for the time-dependence of the change Δx in S is therefore:

$$\Delta x = F_x\, \Delta t^2/2Q\gamma^2\mu. \tag{122}$$

Eq. (95) for the scaling of distance requires that the value in S′ co-moving with the electron must be $\Delta x' = Q^{-1}\Delta x$. The corresponding equation of motion is therefore:

$$F_x' = \mu\gamma^2 a_x' = \mu\gamma^2 d^2x'/dt'^2 = F_x, \tag{123}$$

from which it follows from eq. (115) that

$$\Delta x' = F_x\, \Delta t'^2/2\gamma^2\mu = F_x\, \Delta t^2/2Q^2\gamma^2\mu = Q^{-1}\Delta x, \tag{124}$$

in agreement with expectations based on eq. (95). The scaling is consistent, with elapsed times, distances and inertial masses changing in the same proportion and forces and relative speeds being invariant [note that acceleration scales as $\mathbf{a}' = Q\mathbf{a}$, in agreement with eq. (104)].

There is a separate scaling of properties connected with the gravitational red shift. It is based on a quantity S that plays essentially the same role when the gravitational potential of the observer and/or object of measurement changes that Q plays when the state of motion is varied. The general topic of gravitational scaling is taken up in Chapter 10.

Finally, it is important to return to the question of how electric charges vary between different rest frames. The traditional assumption used by Einstein[5] in discussing the transformation of the Maxwell equations (see Chapter 8. A) has been to assume that charge is invariant, but this position overlooks an important detail concerning the way this quantity is defined in the treatment of electromagnetic interactions. The only essential requirement is that the ratio e^2/ε_0 in Coulomb's Law varies as Q^2, since otherwise electric forces will not be invariant because of eq. (95) for the scaling of distance. This can be arranged in a variety of ways[72] without insisting that charge be invariant, as discussed in Appendix D.

D. Co-moving with the Accelerated Electron

The ALT version of relativity theory differs significantly from Einstein's STR in the way the RP is applied to describe electromagnetic interactions, and this distinction will be considered in more detail in the following. As discussed in the Introduction, it was the view of physicists in the late 19th century[4] that Maxwell's equations (or some variant thereof) must have the same form in different inertial systems. The conclusion was that the way to accomplish this objective theoretically was to find a space-time transformation that leaves such equations invariant. There is an important underlying assumption about the interpretation of the velocity **v** that appears in the Lorentz force equations[71] in eq. (113), however, namely that this quantity varies with the state of motion of the observer. As a consequence, two observers of the same electromagnetic interaction will obtain different results for the transverse force components ($F_{y'z}$), but not for the corresponding value of the parallel component (F_x) since v does not appear in the latter relation.

Einstein used this interpretation in deriving eq. (111) by applying it to the special case in which the observer is always in the rest frame S† of an electron subjected to the Lorentz force. It has already been noted in Chapter 8. A that this choice is at odds with his assumption of the RP since the electron is constantly accelerating and therefore the latter rest frame is not inertial. There is a more specific point that needs to be considered in this case, however. Since $v = 0$ for the electron at all times, the Lorentz force measured by the observer according to eq. (113) in the rest frame of S† would be constant. The electron therefore must move in a constant straight line. For example, if $E_y = E_z = 0$, the electron will continue to move along the x-axis from the vantage point of the S† observer. Yet if there is an external magnetic field with non-vanishing values of the transverse field components B_y and/or B_z, the observer at the origin of the interaction (ORS) will find that the electron veers away from its original direction. The amount of this deviation is proportional to v for the stationary observer in the ORS and thus the direction in which the electron moves can eventually be transverse to the x direction. The question is thus: *Is it reasonable to expect that the direction of the electron's trajectory differs so markedly from the different vantage points of the two observers in S† and the ORS?*

There is another interpretation of the RP in this experiment, however, one which insists upon absolute objectivity in the measurement process. In this view, the quantity **v** in the Lorentz force equations is always measured relative to the origin of the interaction, regardless of the state of motion of the observer. The fixed point in space where the force is applied serves as an ORS for the interaction, so a similar situation exists with regard to clock rates and the UTDL of eq. (57). The speed of the observer relative to the object of the measurement is not directly relevant in this determination, as discussed in Chapter 5. C. First, one has to measure the speed v_{10} of the object relative to the ORS in order to compute the fractional amount of time dilation in the object's rest frame. Then one has to know the speed v_{20} of the observer relative to the ORS to compute the corresponding fractional rate of time dilation in the observer's rest frame. The relative rate for the object clock in the rest frame of the observer is then determined by computing the *ratio* of the latter two quantities, which is equal to the parameter $Q = \gamma(v_{10})/\gamma(v_{20})$ in the corresponding ALT.

In this view, all observers measure the same absolute values for any conceivable property associated with a given interaction, but in accord with the amended version of the RP discussed in Chapter 6. C, their respective numerical values will differ depending on the system of units employed in each case. In the present case, this means that the ORS observer's values of physical properties must be evaluated using the speed v of the object/electron relative to the ORS in which the original Lorentz force was applied (see also the discussion at the end of Chapter 7. D). The corresponding measured values in rest frames other than the ORS do not depend directly on their speed relative to the electron. Rather, they are obtained by using the conversion factor Q in the ALT connecting the given rest frame S with that of the ORS. It is important to note that Q can take on any value. It is the ratio of the rates of stationary clocks in the ORS to those of their counterparts in S. Consequently, if an elapsed time of Δt is measured in the ORS, the corresponding value in S will be $Q^{-1}\Delta t$. A distance of Δr measured by the ORS observer will be converted to $Q^{-1}\Delta r$, whereas an angular momentum $\mathbf{p}_\Theta$ measured in the ORS will be converted to $Q^{-2}\mathbf{p}_\Theta$ [see eq. (98) and Table 1] for the stationary observer in S. This means that all observers, including one who is co-moving with the electron, will measure the same values for each of the three Lorentz force components as the observer in the ORS (since the relevant conversion factor is unity for force components; see Table 1). This is in agreement with the RP and also with the PRM discussed in Chapter 4. A.

In particular, Einstein's observer in S† in the above example [with $Q = \gamma(v)$] finds that the electron veers off in the transverse direction just as does his counterpart in the ORS rest frame where the electromotive force has been applied. No space-time transformation connecting rest frames is required to reach this conclusion. Removing any requirement that the electromagnetic force equations be invariant to a particular transformation leaves open the possibility that any other formulation such as the Hertzian or neo-Hertzian alternative to Maxwell's equations may also be valid.[31] The only essential criterion for any set of such equations is that they agree with experiment in all cases. As a consequence of the above interpretation, there is merit in further amending the RP as follows (see Chapter 6. C):

> The laws of physics are the same in all inertial systems but the units in which they are expressed can and do vary in a systematic manner from one rest frame to another. When a force is applied at a particular location in space, the same results for relevant physical quantities are obtained in all inertial systems after appropriate corrections for differences in standard units have been made. No space-time transformation between different inertial systems is required for this determination.

Chapter 9

Effect of Gravity on Light Waves

The discussion of relativistic effects in Sects. I-VIII has specifically excluded the consideration of gravity. Since it has been found that the properties of objects change when they are accelerated by virtue of an applied force, it is natural to expect that similar changes occur when bodies move through a gravitational field. Until the dawn of the 20th century, physicists were convinced that Newton's Universal Law of Gravitation provides a comprehensive description of the effects of gravity. His inverse square law (ISL) is capable of giving accurate predictions of the motions of planets and stars, and it continues to be used in practical applications to the present day. After Einstein completed his paper on STR in 1905,[5] his attention quickly turned to the broader question of how to incorporate gravitational effects in a consistent manner into the overall theory. He enunciated[38] the Equivalence Principle (EP) as the basis for this extension of kinematic principles. He was guided in this work by his experience with the relativistic theory of electromagnetic interactions and came to the conclusion that, in contrast to the postulate of STR, the speed of light in free space increases with gravitational potential. This conjecture led him in turn to predict that light rays follow a curved path when they pass by massive bodies such as the sun, and that this would lead to observable effects during solar eclipses.

Soldner had reported explicit calculations of the angle of curvature as early as 1801[73] based on Newton's universal theory of gravitation. He based his calculations on Newton's particle theory of light. He found that the results were only slightly dependent on the value for the mass of the particles.[74] His value for the angle of deflection was 0″.84.

Einstein made several attempts to develop a theory of gravitation that was truly relativistic. In a 1911 paper[75] he concluded that clocks run slower in the neighbourhood of massive bodies, and this led him once again to assume that the speed of light must be slower in such regions of low gravitational potential. On this basis he obtained a value of 0″.83 for the angle of light bending near the sun. Unlike Soldner, whose work was unknown to him at the time, Einstein computed this angle by using the wave theory of light and Huygens' principle. At this time he became suspicious of Euclidean geometry and began toying with the idea that space is not flat.[75] Four years later in close collaboration with Grossmann he developed a new theory by making the transition to Riemannian geometry. The resulting General Theory of Relativity (GTR), published in 1916,[76] led Einstein to predict a value for the angle of bending of light rays that was twice as large as his previous result. Two years later it was discovered experimentally[77] that the images of stars were displaced during solar eclipses, and by an amount that agreed within the error of the measurements with Einstein's GTR prediction.

This was a sensational result that cemented Einstein's reputation as one of the great theoretical physicists of all time. The question remains, however, as to whether it actually proves that space is curved, and therefore contradicts Newton's original views on this issue. It has already been shown in the first part of this book that key aspects of STR such as the LT, the ESP, the FLC and Lorentz invariance are not consistent with logical principles, and that all experimental results previously claimed to verify this theory can actually be explained on the basis of an alternative version of the theory based on the VT, the ALT, and especially the uniform scaling of physical properties. This state of affairs at least raises the question of whether there exists an alternative, potentially simpler, version of relativistic gravitational theory than GTR. A significant step in this direction was taken by Schiff in 1960,[78] as will be described in the following section.

A. Assumptions Employed in Schiff's Method

It needs to be kept in mind that there are only three crucial tests of GTR: the gravitational red shift, the deflection of light rays passing close to the sun and other massive bodies, and the precession of the perihelion of Mercury and other planets. Schiff[78] pointed out in the introduction to his paper that GTR's "acceptance rests partly on its conceptual and structural elegance, and partly on its agreement with experimental observation." The stated purpose of his work was to show that the deflection of light rays can be explained in a far less complicated manner than the GTR approach by making a few simple assumptions based on STR and the EP. First, Schiff assumed that local observers at the same gravitational potential always measure the speed of light to have a constant value of c. Secondly, the light is assumed to move in the same perfectly straight-line trajectory for a succession of such observers, that is, the *local* light velocity is always constant in both direction and magnitude. The calculations then proceed on the basis of Einstein's arguments[38,75] mentioned above, namely that the unit of time varies in a well-defined manner [see eq. (4) of Ref.[78]] with the position of the observer in a gravitational field.

However, Schiff made one other key assumption that was not contained in Einstein's 1907 and 1911 papers, and this turns out to be critical in obtaining the desired result for the angle of deflection of light rays. It has a clear relation to the FLC of STR since it states that the unit of distance in the direction *radial* to the sun must vary in inverse proportion to the unit of time, whereas that in transverse direction is independent of gravitational potential [see eqs. (5) and (6) of Ref.[78]], respectively. The FLC is not consistent with time dilation and the light-speed postulate and is therefore invalid. This conclusion has been reached in Chapter 4. B on the basis of the "clock riddle,[32]" but it must be clearly seen that this result is not relevant for the theory of gravitational interactions. The reason is that the FLC of STR is only defined for the case where both the object and the observer are continually located at the same gravitational potential. In the following, Schiff's assumption for distance variations for objects moving through a gravitational field will therefore be referred to as the gravitational FLC or GFLC in order to distinguish it clearly from the FLC derived from Einstein's LT. Note also that Schiff's rule for the rate of a clock is also distinct from conventional time dilation which is induced by a change in its state of motion.

Although the latter approach gives results for the angle of light deflection by the sun in quantitative agreement with Einstein's predictions based on GTR,[76] comparatively little attention was given to Schiff's procedure. Perhaps the main reason for this attitude of contemporary physicists was that a similarly uncomplicated method could not be presented for calculating the other key quantity obtained from Einstein's theory, the relativistic contribution to the precession angle of the perihelion of Mercury's orbit around the sun. This situation has changed recently[79] with the discovery that Schiff's ideas can be applied successfully for this phenomenon by introducing several assumptions regarding the way in which Newton's ISL needs to be implemented in the calculations. On this basis, results are obtained which are in agreement with experimental data for the rate of orbital precession (43".0033/cy calc. vs. 43".2±0.9/cy obs.[80] and Einstein's value of 43".0076/cy,[76] for example). *The same dependence on mean radius and eccentricity of orbit as well as the gravitational mass of the source is found as is the case for GTR*. The corresponding calculations will be discussed in Chapter 11.

The success of the latter approach[79] gives quantitative support to the key assumption on which it is based, namely that all observers, independent of their location in a gravitational field, *who are not in relative motion to one another,* must agree on the distance between any two objects in the universe. This assumption will henceforth be referred to as the gravitational isotropy of space (GIS). On the other hand, as already mentioned, it is assumed that the unit of velocity increases as the object moves to a higher potential, with the component parallel to the gravitational field changing faster than those in the perpendicular direction. It will be seen in what follows, however, that this distinction between the scaling of radial and transverse velocities in Schiff's procedure, *i.e.* the GFLC, is only *a mathematical artifice* whose sole purpose is to accurately compute the *ratio of elapsed times observed locally and at infinity* as the object changes its position in a gravitational field. As such it does not stand in contradiction to the GIS assumption discussed above. Finally, it is also assumed that the units of time and inertial mass scale in the same proportion and thus are both inversely proportional to the unit of perpendicular velocity. By contrast, the gravitational mass of an object is assumed to be the same for all observers (complete details of this gravitational scaling procedure will be given in Chapter 10; see also Ref.[56]). By and large, all of the above assumptions are consistent with the uniform scaling concept discussed in Chapter 7. C except that the details of the scaling are different for changes in the gravitational potential of object and observer than those given in Table 1 for the effects of changes in the state of motion.

One of the main consequences of the GIS is that all observers who are not in relative motion to one another must agree on the path followed by any given object, which in turn must be the same as that measured by a series of local observers analogous to those employed in Schiff's procedure[78] for computing the amount of gravitational deflection of light by the sun. *They will only disagree on the time required to travel a given distance along it.* This result leaves open only one possibility for the actual path followed by the light as it travels from the star to the observer on earth: *it must have travelled in a perfectly straight line.* Why this conclusion is perfectly consistent with the experimental observations made during solar eclipses by Dyson, Eddington and Davidson[77] and subsequently by others,[81,82] as well as with the theoretical calculations of both Einstein[76] and Schiff,[78] is the subject of the following section.

B. Use of Huygens' Principle to Compute the Angle of Deflection

It is an interesting fact of history that a different criterion for computing the angle of deflection of light by gravitational forces was employed by Einstein[76] than was the case for Soldner[73] over a century earlier. In the latter study, Newton's classical theory was applied for an object of small mass that passes close to the sun, and the amount of deflection was obtained as the angle by which the direction of the object's motion changes as it travels between the star and the earth. Both Einstein[76] and Schiff[78] employed a different measure for the amount of curvature, however, one that is based on Huygens' principle. There is a subtle distinction in these two methods. In Soldner's Newtonian approach the direction of the light *velocity* is used explicitly in the calculations, whereas

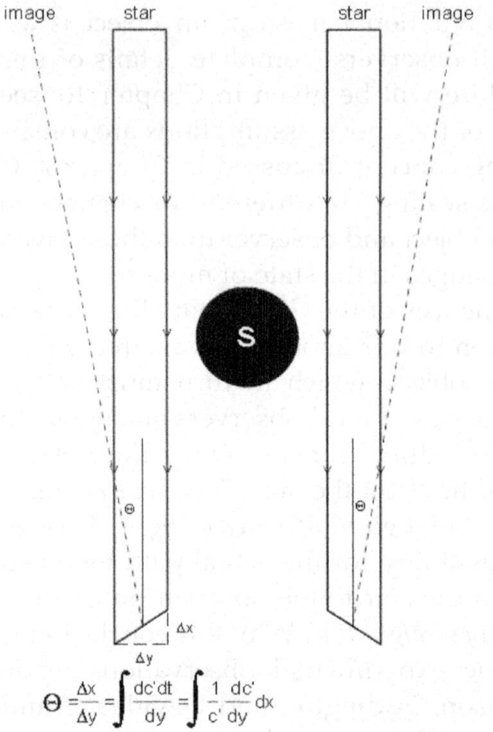

Fig. 4. Schematic diagram showing light rays emitted by stars to follow straight-line trajectories as they pass near the sun. Because of gravitational effects the speed of the light rays c' is known to increase with gravitational potential, with the effect that the corresponding Huygens wave front gradually rotates away from the sun. As discussed in the text, the normal to a given wave front points out the direction from which the light appears to have come, causing the star images to be displaced by an angle Θ during solar eclipses.

Einstein and Schiff only employ the *speed* of the light, not its direction, in arriving at their result.

To see this it is only necessary to examine the formula for the change in angle Θ that they used based on Huygens' principle:

$$d\Theta = 1/c' \, (dc'/dy) \, dx. \tag{125}$$

In this expression dy is the change in lateral distance of the light from the sun in a given time interval, dx is the corresponding total distance travelled, and c' is the speed of the light measured by a stationary observer on earth/infinity over the same infinitesimal period. In Soldner's approach, the path of the light must be curved for there to be a non-zero result for the angle of deflection, whereas in that employed by Einstein and Schiff, all that is required is that the speed of light change with lateral distance from the sun for the above observer.

But can one obtain a non-zero value for dΘ in eq. (125) if the path is not curved? The answer is clearly yes, as the diagram in Fig. 4 shows. To compute the derivative dc'/dy according to Huygens' principle, it is necessary to *compare the speeds of two different light rays* separated laterally by an amount dy. If we assume that the corresponding values of c′ differ by dc′, it is clear that the respective distance travelled in the two cases over time dt will also differ, namely by an amount dc′dt. As shown in Fig. 4, the angle which the line connecting the final positions of the two light rays makes with the corresponding one for their initial positions is thus $d\Theta = dc'dt/dy$. This result in turn can be used to obtain eq. (125) by noting that the total distance travelled is $dx = c'dt$, where c′ is just the average of the two infinitesimally differing light speeds used to obtain the derivative in eq. (125). *There is nothing in this derivation that assumes that either light path is curved, only that the speeds by which the light travels along them are different.*

There is a simple interpretation of this result.[83] The line connecting the current positions of the two light rays simulates a *wave front* in the terminology of Huygens. When the light reaches the observer, the direction from which it has come is judged by extending the normal to this wave front backward in space. Integration of dΘ in eq. (125) over the entire path therefore gives the amount by which the light *appears* to have been deflected from the straight-line path actually followed (Fig. 4). Schiff's work[78] has shown that this angle has a value of 1″.7517 for light coming from infinity which grazes the outer edge of the sun's surface on its way to the earth, identically the same value as obtained by Einstein[76] using GTR. Numerical calculations have also been carried out for this purpose with the (finite time-slice) procedure mentioned above for obtaining the perihelion precession angle for planetary motion,[79] and quantitative agreement is obtained with the latter value on this basis as well. A brief description of these calculations[83] is given below.

Two light rays are assumed to start out at the same distance (10^{12} m in the present numerical example) from the earth, as shown in Fig. 4. The lateral distance between them is Δy, which can be varied in the computer program. Both light rays travel in straight lines along the x-axis and eventually pass by the sun on their way to an observer of the earth's surface. In the specific example under discussion, one of them just grazes the sun's surface, while the other is slightly farther away from it. The origin of the coordinate system employed is placed at the solar midpoint. Since the radius of the sun is taken to be $r_S = 0.696 \times 10^9$ m, this means that the path of the inner light ray is the straight line,

$y = r_S$, while that of the other is $y = r_S + \Delta y$, also a straight line. There is a separate local observer for each of the light rays, following Schiff's procedure.[78] Their trajectories are computed in a series of equal time intervals Δt (0.01 s in the present example) as measured on the stationary clock on earth in each case.

At each step in the calculation the *local* velocity of light is assumed to be c along the x direction, and is resolved into its radial and tangential components relative to the solar midpoint. These are then scaled by different amounts (see Chapter 10. B) in order to obtain the corresponding values in the system of units employed by the observer on earth (actually assumed to be located at an infinite distance from the sun, as in Schiff's original procedure[78]). The corresponding speed of the light (c') from the latter observer's perspective is then computed, which in the present application of the theory is always less than c. It is assumed that the corresponding light ray has travelled a distance c'dt in the –x direction. The new position of the light ray is then computed on this basis and this is used as the starting point for the next time cycle (note that this position is always the same for both the local observer and his counterpart on earth, consistent with the GIS).

Since the speed of light c' is slightly different for the two light rays, it follows that they do not arrive on the earth's surface at the same time. In the present example, with $\Delta y = 1000$ m, the outer ray travels $\Delta x = 0.008492807$ m farther than the one that just grazes the sun's surface (all computations are done in quadruple precision). The line connecting the two end points of the rays has thus rotated by an angle $\Theta = \Delta x/\Delta y$ (see Fig. 4), which is interpreted to be the amount of angular displacement of the star's image. The value obtained for Θ with the above values for Δx and Δy is found[83] to be 1″7517, in excellent agreement with both Einstein's[76] and Schiff's[78] result for this quantity, which in turn is well within the error bars of the measurements of this angle obtained during solar eclipses.[77,78,83]

But if one *assumes* that observers at different distances from the sun will not agree on the path followed by the light, is it not possible that the straight-line trajectory measured locally will appear bent to someone located near the earth? In attempting to answer this question, the first point that should be noted is that the gravitational scaling of time can have no effect on the direction of the light's velocity. Simply employing a faster clock only causes

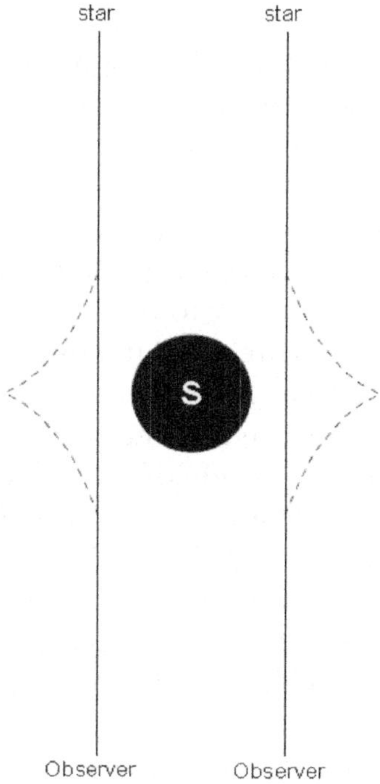

Fig. 5 Diagram illustrating the "pseudo-trajectory" inferred from the velocity vector computed for the observer on earth in Schiff's procedure. Note that on the initial approach the light appears to veer away from the sun (convex trajectory) because the gravitational scaling reduces the magnitude of the radial component relative to the local straight-line path. This result demonstrates that the direction of the latter velocity is *ignored* in Schiff's method, which nonetheless obtains perfect agreement with Einstein's GTR value for the angle of displacement of star images during solar eclipses because of its reliance on Huygens' principle to define this angle (see Fig. 4). Einstein employed the same definition for the displacement angle in his original work.

all velocity components to be decreased in the same proportion, so no change in direction can be expected on this basis. The fact that the radial component of the speed of light is altered more than the transverse component near the sun's surface as a consequence of the gravitational scaling of units does have an effect, but closer analysis shows that *this implies bending in a much different direction than what has previously been inferred from observations of star image displacements*. Since the radial component of the velocity is decreased more for the observer on earth than the corresponding transverse component, he must therefore find a trajectory for the initial part of the light's journey from infinity that is bent away from the sun (see Fig. 5) relative to the straight-line path observed locally (Fig. 4). After the light passes by the solar midpoint, the radial component of its local velocity begins to point away from the sun, however, so that the gravitational scaling implies that the corresponding velocity observed on earth is now directed more toward the sun than is its local straight-line counterpart. The result is a "Δ-shaped" trajectory (Fig. 5) quite unlike the purely concave path around the sun that is normally assumed.

In computing the angle of "curvature" via Huygens'principle, however, the apparent direction of the light velocity from the observer's vantage point on earth is completely immaterial. Instead, only the change in the speed of light with lateral distance from the sun is taken into account.[76,78] Since the light speed increases with distance from the gravitational source, as indicated in Fig. 4, the result is that the wave front of the light is rotated away from the sun, causing the observer to have the *illusion* that the star has been displaced in this direction. In effect, the fact that the *direction* of the light's velocity appears to differ from one observer to another on the basis of the gravitational scaling is *totally ignored in the calculation*; only the magnitude of the velocity (speed) observed on earth is actually considered in applying Huygens' principle. *The corresponding direction is always the same as for the local observer*. Thus the Δ-shaped trajectory obtained in Schiff's treatment[78] can simply be viewed as an artifact of the overall theoretical approach employed to compute the amount of star-image displacement.

The conclusion from the above analysis is that one should carefully distinguish between a change in the direction of the trajectory that light actually follows on the one hand, and the angle of deflection indicated by photographic images of the light source on the other. Just the fact that light speed varies with lateral distance from a gravitational source, which is a clear assumption in Schiff's procedure,[78] is sufficient to lead to the quantitative prediction of the observed displacement of star images during solar eclipses. The path that the light follows cannot be determined from application of Huygens' principle alone, but the assumption employed by Schiff to obtain exactly the same amount of deflection as Einstein did by using GTR, namely that the velocity of light is always the same in both magnitude and direction for a series of local observers, speaks strongly for the conclusion that its trajectory is actually a straight line. *The effect of the sun's gravitational attraction is merely to slow the light down, but not to cause the individual photons to alter their direction.*

C. Experimental Test of the Straight-line Interpretation

It is clear from the outset that it is a practical impossibility to measure light paths with sufficient accuracy to obtain a definitive

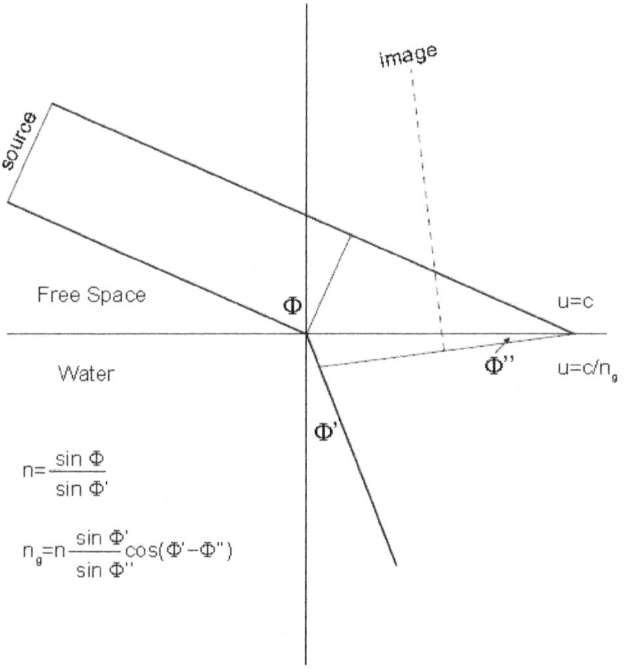

Fig. 6. Angles of incidence Φ and refraction Φ' as light enters water from free space. The corresponding angle Φ" which the wave front makes with the interface upon entering a condensed medium can be determined by noting the direction from which the image of the light source appears to arrive inside the medium. Note that Φ" is generally different than Φ'. As discussed in Chapter 9. C, measurement of all three angles enables the determination of both the normal refractive index n and the corresponding group refractive index n_g of the condensed medium.

answer to the question of whether gravitational forces really deflect light or merely produce the illusion that this has occurred. One must therefore rely on indirect evidence to decide this matter as objectively as possible. One such experiment is outlined below.

The main conclusion of the preceding section is that the shifting of star images during solar eclipses is caused solely by the fact that light travels with different speeds depending on its proximity to the sun. It is possible to make a close analogy with ordinary light refraction on this basis.[84] In conventional experiments the angles of incidence and refraction are measured by *direct observation of the path* of the light as it passes through an interface between two media. The ratio of the sines of these two angles defines the refractive index n in accordance with Snell's law. However, the speed of light v is not determined by n alone but rather by n_g, the group refractive index, specifically $v = c/n_g$ and not c/n.

The diagram given in Fig. 6 illustrates the situation in detail for light entering from free space with an angle of incidence Φ relative to the normal to the interface of a condensed medium of refractive index n. The corresponding angle of refraction is Φ′ such that

$$\sin(\Phi') = \sin(\Phi)/n. \quad (126)$$

The light ray on the left of the diagram enters the dispersive medium first, at which point its speed changes from c to c/n_g [see eq. (33) and the ensuing discussion]. There is a delay before the light ray on the right enters the medium, during which time it continues to move with speed c. After both light rays have entered the condensed medium, the line connecting them makes an angle Φ″ with the interface, but this is generally not the same as Φ′ (Fig. 6). Using trigonometric identities one finds that

$$n_g = n\,(\sin\Phi'/\sin\Phi'')\cos(\Phi' - \Phi'') \quad (127)$$

$$\cot(\Phi'') = n_g/(n \sin\Phi' \cos\Phi') - \tan\Phi'. \quad (128)$$

From these relations it is clear from eq. (127) that the condition for Φ′ to be equal to Φ″ is that $n_g = n$, which according to eq. (33) can only occur if the refractive index does not change with frequency.

According to the arguments of Chapter 9. B, the image of the light source *when viewed from within the dispersive medium* will appear to lie along the normal to the wave front, *i.e.* at an angle of Φ″ relative to the interface. Measurement of this angle will therefore give a result that is generally different than the angle of refraction (Φ′) obtained by observing the light path from outside the medium. The above equations indicate that one can determine n_g on the basis of separate measurements of both these angles. The result should be the same as obtained for n_g on the basis of measurements of n at a series of light frequencies, which in turn should also agree with the value obtained by explicitly measuring the speed of light of the same frequency in the dispersive medium.[20,85,86]

In the present context, however, the most significant result of such an experiment would be to show that the angle of deflection of the light image is generally *not the same* as the angle by which the light trajectory itself deviates from its initial direction. It would therefore lend strong support to the conclusion based on Schiff's method[78] for computing the angle of gravitational deflection by the sun that the path followed by the light on its way from the star to the observer on earth *is actually a perfectly straight line.*

D. Relationship to Black Holes

Even before Soldner[73] published his calculations on the gravitational bending of light, there was speculation[87] by Michell that an object might be so massive that it would become impossible for light to escape from its surface. As discussed above, the argumentation in GTR is fundamentally different than in the Newtonian approach to gravity, but the belief still persists that such "black holes" exist and that they do not allow light to pass from them. Hawking[88] has argued that high-energy radiation can still escape from the surface of a black hole, however.

In the previous sections it has been shown that all known experiments regarding the phenomenon of gravitational light deflection can be explained quantitatively by assuming that light always travels in a perfectly straight line. It is therefore of interest to see how the theory of black holes is affected by making this assumption. First of all, it should be noted that this position is still consistent with Newton's ISL provided that one takes account of the fact that the acceleration due to gravity from the g field on an object varies with the state of motion of the observer. Ascoli[89] has concluded that when an object is moving with speed v relative to the local observer, its acceleration due to gravity is damped by a factor of $\gamma^{-2} = 1 - v^2c^{-2}$. A consistent derivation of this result will be given in Chapter 10. D. This relation has been used successfully in the calculations mentioned above for the precession angle of Mercury's perihelion.[79] In the case of light, for which the local value of v is always c, *this damping factor is exactly zero, so that no acceleration is to be expected.* Thus this result is consistent with both Schiff's approach[78] and the underlying theory of the calculations of Mercury's orbit.

According to this view, light can pass as closely as possible to the surface of a black hole without being deflected. The apparent shift in the position of the image of the light source will be very much larger than it is for the sun, however. Moreover, there is no gravitational effect keeping light from escaping the interior of a black hole, so γ rays are expected to be observed, and not only those originating outside the boundary of the black hole. It should not be forgotten thereby that there is a quite high probability for photons to be absorbed because of the high density of matter, however, so on this basis the description as a blackbody is certainly appropriate. It is also clear that the speed of light will be quite small in the interior of a black hole because of the gravitational time dilation, and a very large red shift for light escaping from it is also expected for an observer located at a relatively high gravitational potential. The key point remains, however, that none of these effects need involve true gravitational deflection, as they are all consistent with a perfectly straight-line trajectory. The phenomenon of gravitational lensing is also expected on this basis, provided the light source is located directly behind the black hole. The image of the light source would be significantly distorted relative to that which would be detected in the absence of the black hole (see Fig. 4).

E. Summary

The present analysis and explicit calculations indicate that one must draw a clear distinction between the angle of deflection of a star's image and the corresponding angle by which the emitted light deviates from its initial direction. Huygens' principle, which is used to define the angle of deflection in GTR, gives no information about the actual path the light follows. In order for "curvature" to be observed on this basis, it is only necessary for adjacent light rays to travel at different speeds, something that is known to occur near massive objects. This effect causes the wave front of the light emanating from the star to be rotated away from the sun. Simply demonstrating that the images of stars are displaced during a solar eclipse therefore tells us nothing about the actual path taken by the light on the way to the observer.

The work of Schiff[78] has demonstrated that one can obtain the same value for the deflection angle as in GTR by assuming that the trajectory of the light is actually a straight line. The numerical calculations carried out in the present study show clearly that the wave front of light emitted by a star is rotated away from the sun by exactly the latter angle, causing the star's image to appear to be displaced by this amount (Fig. 4). Since it is impossible to actually observe the path taken by the light from such a distant source, there is no direct way of either verifying or denying the hypothesis that it always travels in a straight line. Nonetheless, an experiment has been outlined for testing the conclusion that the angle of displacement of the image of a light source is not the same as the angle of refraction in dispersive media. Such a procedure should make it possible to measure the group refractive index n_g directly, instead of relying on a series of measurements of the corresponding refractive index n for different frequencies of light, as is normally done for this purpose.

The underlying basis for the present calculations of gravitational interactions is the fact that the units of time and velocity differ from one gravitational potential to another. Nonetheless, it is assumed in accord with the GIS (see Chapter 9. A) that the distance between any two objects is the same for all observers in the universe as long as they are not in relative motion to one another, independent of both the position of the objects and that of the observer in a gravitational field. The same relationship is assumed for gravitational masses. Following Schiff,[78] however, the unit of velocity is assumed to vary with gravitational potential, and to be different for the component radial to the field than for that perpendicular to it. Analysis shows, however, that the latter distinction is only an artifact of Schiff's method whose sole purpose is to compute the ratio of elapsed times observed locally and at infinity (the earth), respectively. It leads to a value for the velocity of light that has the correct magnitude but the wrong direction (see Fig. 5). The true direction is the same as for the local observer, which in turn is obtained by the reverse scaling procedure in Schiff's method. As a consequence, all observers *not in relative motion to one another* agree completely on the path taken by the light. Once the speed of light for an observer at infinity is known, it is easy to compute its value for any other observer not moving with respect to the first simply by taking account of the difference in their respective clock rates. The appropriate scaling factor required to make this conversion will be discussed in Chapter 10.

In the present application, the velocity of light is assumed to be constant in both magnitude and direction, also following Schiff. A connection has been noted between this assumption and earlier work of Ascoli.[89] When applied to optics, the latter result leads to the conclusion that photons are not accelerated by gravitational forces and therefore must always travel in a perfectly straight line according to Newton's First Law. This result is also consistent with Newton's Third Law because it indicates that the trajectories of photons are not affected by gravitational forces on the one hand, and on the other, that their null gravitational mass precludes their exerting a gravitational force on any other object.

Finally, the computational method[83] employed to compute the angular displacement of star images during solar eclipses has also been extended to the other key application in Einstein's original work, calculation of the relativistic contribution to the precession angle of the perihelion of planetary orbits. This objective was discussed in Schiff's original work,[78] but apparently was never achieved by him. Quantitative agreement with Einstein's closed expression for the perihelion precession angle is obtained with the present computational approach,[79] however, thereby giving considerable support to its underlying assumptions.

Chapter 10

Equivalence Principle and Gravitational Scaling

The calculation of the angle of light deflection caused by the sun and massive bodies in general is based to a large degree on Einstein's EP.[38] However, in Chapter 9. A it has been noted that Schiff[78] made an additional key assumption about length variations (the GFLC) that goes beyond the prescriptions of the EP. Without this assumption, the same result for the angle of deflection follows from Schiff's procedure as from Einstein's pre-GTR value,[75] which is 50% smaller than what is now accepted to be correct. It is well at this point to emphasize that the EP and the various scaling relationships discussed above are essentially *ad hoc* in nature. Ultimately, all such assumptions/postulates are judged on their ability to predict the results of experiments in a consistent manner. The same characterization applies to Newton's Laws of Kinematics and the Laws of Thermodynamics. However, it needs to be kept in mind that GTR also rests on various assumptions and therefore has a definite *ad hoc* character as well. There are no "first principles" to rely on at this level of theory. No matter how elegant and mathematically sophisticated a given physical theory is, it must be declared to be invalid as soon as any piece of incontrovertible evidence emerges that contradicts its predictions.

This was the fate of the LT, for example. In this case the evidence in question is the contradiction that emerges from the clock riddle[32] discussed in Chapter 4. B. It is unacceptable that a theory makes different predictions for the same quantity depending on how it is applied. However, by eliminating the offending assumption in Einstein's STR,[5] the revised version of the theory attains an objective character that was notably lacking in the original. It subscribes to a uniform scaling principle that turns out to depend on a single parameter, as indicated by the results in Table 1. Einstein's EP is objective in character, so there is a clear possibility for finding common ground on this basis. The following discussion addresses this possibility by developing a gravitational scaling procedure analogous to that discussed above for objects that are subjected to an applied force (kinetic scaling).

A. Variation of the Unit of Energy

The classical view of how the total energy of an object varies with altitude is that there exists a gravitational potential energy. When the object falls it is subject to a constant acceleration, which according to Galileo is exactly the same for all masses (unicity principle), but its energy does not change due to the fact that the accompanying increase in kinetic energy is exactly balanced by a decrease in its potential energy. Einstein had shown[5] that the energy of an object at rest is equal to the product of its inertial mass m_I and the square of the speed of light ($E = m_I c^2$), as discussed in Chapter 7. A [see eq. (80)]. He deviated further from the traditional view by assuming that the increase in kinetic energy as the object falls from point P to point O is due to the fact that *the unit of energy decreases* with its distance dr from a gravitational source. He concluded that the relationship between the respective total energies at these two points is given by the formula:

$$E_O = E_P\,(1 + gdrc^{-2}), \qquad (129)$$

where $dr > 0$ is the (differential) vertical distance between the two points and g is the local acceleration due to gravity as computed by Newton's Universal Law of Gravitation. This conclusion is clearly in line with the PRM of Chapter 4. A and the uniform scaling procedure (see Chapter 6.C) employed in the ALT version of special relativity.

When an object falls a finite distance from infinity toward a gravitational source, the corresponding change in the unit of energy is obtained by integration[5] of the quantity in parentheses on the right-hand side of eq. (129). It is helpful therefore to make the following definition:

$$A_O = 1 + G\,M_S/\,c^2\,r_O, \tag{130}$$

in which G is the universal gravitational constant (6.67×10^{-11} N m^2/kg^2), M_S is the gravitational mass m_G of the source (active mass) and r_O is the distance from point O to the source. The unit of energy is A_O times *smaller* at O than it is at an infinite distance from the gravitational source. The quantity A_O plays an analogous role in the gravitational scaling procedure as the UTDL γ factor does for kinetic scaling in Table 1: Clocks at rest in the ORS [see eq. (57)] run $\gamma(v_{10})$ times faster than their counterparts in the rest frame of inertial system S moving at speed v_{10} relative to the ORS. In other words, a location at infinity plays the analogous role for gravitational scaling as the ORS does for kinetic scaling. In that sense it is easier to apply because there is never any question about the reference system for computing the value of the appropriate scaling factor.

As a consequence of the above definition, it follows that the ratio of the energy values measured for the same object by two observers located at different distances from the gravitational source, O at r_O and P at r_P, is given by

$$E\,(O) = (A_O/A_P)\,E(P) = S\,E(P) \tag{131}$$

If O is located at a higher altitude than P, *i.e.*, if $r_O > r_P$, his energy value will be smaller than that of P because $A_O < A_P$ according to eq. (130). In the following discussion the ratio A_O/A_P will be referred to as S. It is then analogous to the ALT clock-rate ratio Q employed in kinetic scaling. It will be shown that the ratios of the respective measured values by observers O and P for all physical quantities are given by integral powers of S, just as in kinetic scaling the analogous ratios are always integral powers of Q given in Table 1. An important goal of the overall relativistic theory thus becomes the determination of the powers of Q and S, respectively, that correspond to each physical quantity such as energy, time and distance.

The conservation of energy principle allows one to obtain a relationship between A_O in eq. (130) and γ in eq. (57). As an object falls from P to O, its speed, and therefore its energy in terms of the local units, increases. If the speed of the object (measured by an observer who is in a gravity-free region of space and is at rest relative to the active mass[78]) is u_P at P and u_O at O ($u_O > u_P$), its energy *increases* as the ratio $\gamma(u_O)/\gamma(u_P)$ during the fall if gravitational effects are ignored. On the other hand, according to eq. (131) its energy is *lower* when it reaches O by a factor of $S = A_O/A_P$ if kinetic effects are not taken into account. Since energy is actually conserved throughout the fall, it therefore follows that

$$\gamma(u_P)/\gamma(u_O) = A_P/A_O, \quad (132)$$

i.e., $\gamma(u_P)$ increases in direct proportion to A_P as the object falls in the gravitational field. Schiff employs a similar argument[78] to obtain the equivalent of eq. (129) based on a power-series expansion of $\gamma(u)$ and another relation from Newton's classical gravitational theory. One may look upon the proportionality in eq. (132) as the definition of the A_O and A_P factors, which in turn may be slightly different than that given in eq. (130) in actual numerical calculations to be discussed below.

B. Gravitational Scaling of Time, Length and Mass

In his 1907 paper[38] Einstein used the EP to prove that clocks run faster as their distance from a gravitational source increases. The relationship between elapsed times measured by two observers at different locations in the gravitational field is

$$T(O) = (A_P/A_O)\, T(P) = S^{-1}\, T(P). \quad (133)$$

Measurements of the rates of standard atomic clocks have verified Einstein's prediction quantitatively.[90]

Comparison of eq. (133) with eq. (131) reveals something quite interesting, however. The factor in parentheses in eq. (133) for time is the reciprocal of that for the ratio of energy values. The power of $S = A_O/A_P$ is + 1 for energy by definition, but it is equal to –1 for time. This is in stark contrast to the situation found in the case of kinetic scaling of the same two quantities. In that case the power of Q in Table 1 is the same for both, namely + 1. This means, for example, that when the speed of a meta-stable object increases relative to its ORS, both its energy and its lifetime become greater. On the other hand, when the same object is taken to a higher altitude, its energy increases but its lifetime *decreases* by the same factor. It is especially interesting that Einstein was able to correctly predict the variation of radiative periods with distance from a gravitational source by using arguments that are based directly on STR,[5] that is, by means of his EP. Yet energy and lifetime vary in opposite directions when the position of the object in a gravitational field is changed, but in the same direction when their speed relative to the ORS is altered. When the metastable object is in free fall from P to O, its energy stays constant so that eq. (132) holds, but its lifetime increases by the square of $S = A_O/A_P > 1$. The rates of clocks decrease because of the increased speed as a consequence of the fall, but they also decrease in the same proportion by virtue of the change in altitude, so that the overall effect is magnified by S^2.

The above scaling relationships also have an interesting application with regard to Planck's radiation law. As already discussed in Chapter 7. C, the ratio of the energy and frequency of emitted radiation varies with the state of motion of the source relative to the observer. This is most easily seen from the fact that the unit for Planck's constant h is J s. As the speed of the radiative source increases, *both* the energy and time units increase as well. According to the RP, the value of h is the same in every inertial system, but only when expressed in terms of the "local" units. Since both the energy and time units increase in direct proportion to Q, it follows that the energy/frequency ratio *of a moving object* measured by the stationary observer varies as Q^2. For him the energy of the radiation increases while the frequency decreases by virtue of the transverse Doppler effect. Since the units of energy and time are inversely proportional in gravitational scaling, however, no such variation in the energy/frequency ratio occurs when only the position of the source in the gravitational field is changed relative to the observer. Indeed, this fact was paramount in Einstein's original conclusion that clock rates increase with distance from the gravitational source.[38]

The question remains as to how the unit of length scales with position in a gravitational field. Schiff[78] assumed in his calculation of the angle of deflection of light passing near the sun (Chapter 9. A) along a direction *perpendicular* to the field are independent of the distance from the gravitational source, while those measured in a radial direction are proportional to S. He refers to a specific observer in this case, however, namely one who is located at infinity and is stationary in the solar rest frame. In other words, such an observer finds that an object "grows" in the radial direction as it moves to higher altitude, whereas no change in directions perpendicular to the field would be noted. As a consequence, the unit of distance would seemingly have to vary with orientation in the gravitational field.

A key point that is easily missed in discussing Schiff's method, however, is that at each stage of the calculation the starting point is a local straight-line trajectory for the light. To obtain the correct result for the displacement angle it is necessary to always continue the scaling procedure from a point on the local trajectory. On this basis the only way to have internal consistency in this approach is to assume that *both* the local observer and his counterpart at infinity measure *the same* trajectory for the light, in accordance with the GIS of Chapter 9. A. The fact that his method leads to the correct value for the displacement angle can be explained by assuming *ad hoc* that the radial distance scaling in his treatment only applies to the component of velocity v_r * in that direction *when measured from the vantage point of an observer O* in a gravity-free region of space who is stationary in the rest frame of the sun.* In other words,

$$v_r^*(O^*) = (A_O/A_P)^2 v_r(P) = S^2 v_r(P). \tag{134}$$

The corresponding scaling relationship for the tangential component v_t is

$$v_t(O^*) = (A_O/A_P) v_t(P) = S v_t(P) \tag{135}$$

The additional factor of S in eq. (134) does *not* imply that the component of the distance vector between two points that lies in the direction radial to the gravitational field of the sun is not actually the same for both observers. Note that there is a common factor of S in eqs. (134-135) due to the scaling of time, as indicated in eq. (133).

In this connection it is helpful to define a separate quantity L_r^* to be distinguished from the actual radial component of the distance L_r. They scale differently, namely as

$$L_r\,(O) = L_r\,(P), \tag{136}$$

$$L_r^*\,(O) = (A_O/A_P)\, L_r\,(P) = S\, L_r\,(P). \tag{137}$$

There is no need to distinguish between corresponding transverse distance components (L_t), *i.e.* they satisfy eq. (136) directly. In practice in Schiff's computational procedure,[78] eq. (137) is used by the observer at infinity (O*) *only* to determine the radial component of the velocity of light from his perspective.[83] In order to simulate this procedure in a computer program it is necessary to determine both the velocities **v** (O*) and **v** (P). The direction followed by the light is parallel to **v** (P) for both observers, consistent with the above arguments.[83] The path taken is a straight line in each case *and is exactly the same for both observers*. The *speed* with which the light follows this trajectory is different for the two observers, however. It is determined by the absolute values of **v** (O*) and **v** (P), respectively. In effect, the scaling of the radial component of the speed of light only affects the elapsed time required by the light to follow a particular trajectory as measured on a given observer's clock. More details regarding the above calculations and their interpretation are given elsewhere.[83]

Experience with calculations of planetary trajectories[79] indicates the scaling of *gravitational* mass m_G is especially simple by contrast. One obtains consistent results from Newton's classical theory, for example, by assuming that the gravitational masses of the sun and planets are independent of both their state of motion relative to the observer and their location in a gravitational field. Hence,

$$m_G\,(O) = m_G\,(P). \tag{138}$$

This scaling assumption has been employed successfully in calculations of the precession angle of Mercury's perihelion.[79]

As noted in eq. (94), the unit of inertial mass m_I of an object does change with its state of motion, however. The same is true for gravitational scaling, but the corresponding power of S is not the same as for Q in kinetic scaling. The unit of inertial mass is N s^2m^{-1}, which is to be distinguished from kg, the unit of gravitational mass m_G. Accordingly, m_I scales as Q but as S^{-1} because of the way that energy, time and length vary with gravitational potential [see eqs. (131,133,136)]. A summary of the factors involved in kinetic and gravitational scaling is given in Table 2. Their application to the computation of trajectories of objects in gravitational fields is the subject of Chapter 11. It was also used in Chapter 9. B to compute the deflection angle of light.

Table 2. Variation of the units of various physical quantities with gravitational potential. The object of the measurement is at point P in the gravitational field, whereas the observer carrying out the measurement is at point O. The ratio of the respective units of a given quantity is conveniently given as the integral power n of the ratio $S = A_O/A_P$ [see eq. (130) for definition]. The corresponding ratio in kinetic scaling is indicated in the last column by means of the integral power q of the ratio Q (see Table 1) in each case.

Physical Quantity	Standard Unit (mks system)	Power of ratio S n	Power of ratio Q q
Energy	J = N m	1	1
Time	s	–1	1
Length	m	0	1
Gravitational mass	kg	0	0
Velocity	m/s	1	0
Differential length[a] radial to field (L_r*)	m*	1	1
Velocity radial to field[a]	m*/s	2	0
Inertial mass	N s^2/m	–1	1
Acceleration	m/s^2	2	–1
Accel. due to gravity g[a]	kg*/m*^{2}, m/s^2	–3	–2
Speed of light c	m/s	1	0
Force	N	1	0
Angular momentum	N m s	0	2
Momentum	N s	0	1

[a] Scaling only used in Schiff's procedure[78] with $A_o = 1$ (observer O* at infinity)

C. The Pound-Snider-Rebka Experiment

The gravitational red shift derived on the basis of the EP in Einstein's Jahrbuch review[38] is expressed in the equation below [note that it is equivalent to eq. (133) with X = P and D = O after using the definition in eq. (130) and eliminating insignificant terms]:

$$\nu_D = \nu_X(1 + \Phi c^{-2}). \tag{139}$$

In this formula, ν_X is the frequency of light emitted from a source that is at rest at a given gravitational potential, whereas ν_D is the corresponding value of the frequency detected by an observer who is not in relative motion to the source but is located at some other position in the gravitational field. The potential energy difference Φ between the source and detector determines the fractional amount of the frequency shift. For an experiment near the earth's surface, $\Phi = gd$, where d is the difference in altitude between the source and detector and g is the acceleration due to gravity. The sign convention is such that $\Phi > 0$ when the light source is located at a higher gravitational potential than the detector.

The same proportionality factor occurs[38,75] in the relationship between the light speeds measured at the same two positions in the gravitational field, where c_D is the value measured at the detector and c_X is the corresponding value measured at the light source:

$$c_D = c_X\,(1 + \Phi c^{-2}). \tag{140}$$

As a result, it is clear that according to the EP both the frequency and the speed of light increase as the light source moves to a higher gravitational potential. Moreover, the fractional amount of the change is the same in both cases. As will be discussed in the following, there is ample experimental evidence to indicate that both of these relations are correct.

Terrestrial confirmation of the gravitational red shift first became possible with the advent of the Mössbauer technique for detecting changes in x-ray frequencies. Pound and Snider[91] placed a ^{57}Fe source at a distance d of 22.5 m above an absorber. According to eq. (139), the fractional change in the x-ray frequency (3.47×10^{18} Hz) should have a value of $gdc^{-2} = 2.45 \times 10^{-15}$. The authors employed the EP directly to obtain their results. By imparting a downward velocity of $v = -gdc^{-1}$ to the detector, the resulting first-order Doppler effect exactly balanced the gravitational red shift:

$$\nu_D = \nu_X(1 + vc^{-1}). \tag{141}$$

The observed results were found to be within expectations by a factor of 0.9990±0.0076.[91]

The interpretation of the above experiment is nonetheless a matter of some confusion. Pound and Snider[91] basically avoided a theoretical discussion of the ramifications of their work *vis-à-vis* GTR, but they noted that "the description of the effect as an 'apparent weight' of photons is suggestive. The velocity difference predicted is identical to that which a material object would acquire in free fall for a time equal to the time of flight." The latter value is assumed to be $\Delta t = dc^{-1}$, so that the expected velocity increase of the falling particle would indeed be $\Delta v = g\Delta t = gdc^{-1}$. However, it is important to see that the frequency of the emitted light *in absolute terms* is actually the same at the light source as it is on the ground below when it arrives there. As Einstein stated in his 1911 paper,[75] this is a clear physical requirement because the number of wave crests arriving at a given point over a specific time interval is a constant.[92] The reason that the detector at the ground level records a higher frequency is because the unit of time there is greater than at the location of the light source. Therefore, more wave crests are counted in 1 s on the ground clock than for its counterpart at a higher altitude. Comparison of the rates of atomic clocks located at different altitudes over long periods of time has confirmed Einstein's conclusion quantitatively on this point.[90]

The above discussion raises another question, however, namely what happens to the speed of light as it descends from the source? According to eq. (140) and Einstein's EP,[38,75] the answer is clear that it decreases by the same fractional amount as for the associated frequency red shift. To be specific for the Pound-Snider example, the speed of light must have a value of c at ground level for an observer located there, whereas the same observer measures the value at the light source to be $1 + gdc^{-2}$ times larger. In other words, the speed of light actually *decreases* as it passes downward from the source to the detector at ground level. This is clearly the opposite change that is expected from the "apparent weight" argument alluded to above. Again, there is ample experimental evidence that Einstein's eq. (140) is quantitatively accurate. One knows, for example, from the work of Shapiro[93] that radio signals slow in the neighborhood of the sun as they approach massive bodies such as Venus and Mercury. There are time delays between the emission of radio pulses towards either of these planets and detection of their echoes when they return to the earth's surface. This shows that the speed of light decreases under the influence of a gravitational field, *i.e.* with $\Phi < 0$ in eq. (140).

There is another important question that needs to be considered in the present discussion, however, namely what happens to the *wavelength of light* as the source changes its position in a gravitational field? As mentioned above, according to Einstein,[38] "light coming from the solar surface has a longer wavelength than the light generated terrestrially from the same material on earth.[94]" Since this statement comes directly after his proof of eq. (139) for frequencies, it seems quite plausible he was basing it on the assumption that the product of wavelength λ and frequency ν is equal to the speed of light. In order to obtain the quoted result on this basis, however, it was also necessary for him to assume that the speed of light is actually the same at the sun's surface as on the earth. This explanation is nonetheless in conflict with eq. (140), which it must be emphasized appears in the same publication.[38] Making the same assumption about the relationship between light speed, wavelength and frequency and comparing eq. (140) with eq. (139) leads unequivocally to the conclusion that the wavelength of light for a given source is completely independent of the latter's position in a gravitational field:

$$\lambda_D = c_D \nu_D^{-1} = c_X \nu_X^{-1} = \lambda_X. \tag{142}$$

It is important in general to recognize that in each of eqs. (139,140,142), the object of measurement is the same for two different observers. The only reason why these observers do not obtain the same result in each case is because they employ different units for their various measurements. The quantity in parentheses in eqs, (139,140), $S = 1 + \Phi c^{-2}$, can be looked upon as a *conversion factor* between their respective units. As shown in Table 2, the conversion factors for other physical quantities can always be expressed as powers of S, just as the corresponding kinetic scaling factors are always powers of Q. As before with the kinetic scaling, the gravitational conversion factors satisfy the ordinary rules of algebra. For example, since frequency and time are inversely related, it follows from Table 2 that the conversion factor for time is S^{-1}. The unit of velocity or speed is equal to the ratio of the distance travelled to elapsed time and hence the conversion factor in this case is equal to the ratio of the corresponding conversion factors, namely $S^0/S^{-1} = S$, consistent with the previous definition from eq. (140).

It is useful to employ the above concepts to follow the course of the light pulses in the Pound-Snider experiment.[91] The analysis of two events is sufficient to illustrate the main points in this discussion, namely the initial emission of the x-rays at the higher gravitational potential (I) and their subsequent arrival at the absorber (II). It is important to use the same set of units in comparing the corresponding measured results. In terms of the units at the x-ray source, the observer there measures the following values for event I: ν_X (I) = ν,λ_X(I) = λ and c_X(I) = c, where ν and λ are the standard frequency and wavelength of the light source. Upon arrival at the absorber, the same observer measures the corresponding values: ν_X(II) = ν,λ_X(II) = $S^{-1}\lambda$ and c_X(II) = $S^{-1}c$. In other words, the value of the frequency has not changed during the passage of the x-rays down to the absorber, but the speed of light has decreased in accordance with the EP ($S > 1$). Consequently, the wavelength of the light must have decreased by the same factor in order to satisfy the general relation (phase velocity of light) in free space between frequency, wavelength and light speed ($\lambda_X \nu_X = c_X$).

The observer located at the absorber measures generally different values for the same two events, not because he is considering fundamentally different processes, but rather because he bases his numerical results on a different system of physical units (PRM of Chapter 4. A). He therefore finds for event II: ν_D(II) = $S\nu$, λ_D(II) = $S^{-1}\lambda$ and c_D(II) = c. His unit of distance is the same and so there is complete agreement on the value of the wavelength ($S^{-1}\lambda$). However, his units of frequency and speed are S times *smaller* than for his counterpart at the higher gravitational potential, and therefore he measures values for these two quantities which are S times *greater* in each case. This set of results illustrates the important general principle of measurement, namely *that the numerical value obtained is inversely proportional to the unit in which it is expressed.* The corresponding values for the initial emission process are accordingly: ν_D (I) = $S\nu$, λ_D(I) = λ and c_D(I) = Sc.

In absolute terms what has happened as a result of the downward passage of the x-rays between source and detector is that the light frequency has remained constant throughout the entire process, exactly as Einstein argued in his Jahrbuch review.[38] On the other hand, both the light speed and the corresponding wavelength of the radiation have deceased by a factor of S. This is in agreement with Einstein's second postulate of STR[5], namely that the observer at the absorber must find that the speed of light when it arrives there has a value of c in his units because it is at the same gravitational potential as the observer at that point in time.

Values of linear momentum $\mathbf{p} = m_I\mathbf{v}$ are the same for observers located at different gravitational potentials, since the corresponding scaling/conversion factors for inertial mass and velocity exactly cancel one another. With reference to the Pound-Snider experiment,[91] this means that x-ray photons increase in momentum as they fall through the gravitational field. This follows from quantum mechanics and the fact that the corresponding wavelength of the radiation decreases in the process. Hence, the $E = pc$ relation holds for observers located at all gravitational potentials since the light speed is decreasing at exactly the same rate as the momentum is increasing, while the energy of the photons is conserved throughout the entire process. This situation might seem counter-intuitive, but it emphasizes that photons are different than material particles with non-zero rest mass, for which momentum and velocity are strictly proportional (see also the related discussion of light refraction in Ref. [95]). Note also that because of the governing quantum mechanical equations, this result is consistent with the $\nu\lambda = c$ relation for light waves, also as already alluded to above.

D. Scaling Factor for g

The main thesis in Ref.[78] is that the trajectories of objects moving under the influence of a gravitational field can be accurately computed using the methods of Newtonian mechanics when account is taken of the way the units of physical quantities change with distance from the active mass. Schiff speculated that the same techniques he employed to compute the angle of displacement of star images during solar eclipses could be successfully applied to the determination of the orbital trajectories of planets such as Mercury and Venus. In order to accomplish this objective he claimed that it was necessary to have "an equation of motion for a particle of finite rest mass" and to take account of the fact that the planets move at speeds less than c. He also felt that higher orders of the $S = A_O/A_P$ factor [see eq. (130)] would have to be included in such calculations.

In a more recent study, Schiff's approach has been modified successfully[79] to obtain the angle of precession of the perihelion of the orbits of Mercury and other planets. Rather than relying on an equation of motion for the planets, it is simply assumed that

Newton's Universal Law of Gravitation holds for a *local observer* M, *i.e.* for someone at the same distance r from the sun and co-moving with the planet. Accordingly, the local acceleration to gravity is given by Newton's ISL as:

$$a_M = g = GM_S\, r^{-2}, \tag{143}$$

In order to carry out the calculations from the perspective of an observer at infinity (O*), it is necessary to know how g scales with both the relative speed of the planet and its distance from the sun. For this purpose it is important to follow Schiff's suggestion[78] of introducing a third observer P who is located at the same gravitational potential as M but is stationary with respect to O* and the sun.

For the kinetic scaling (Table 1) it is assumed that g in eq. (143) varies as Q^{-2}.[89] This follows directly from the fact that the distance r scales as Q, whereas the gravitational mass M_s is the same for all observers. In the present case, $Q = \gamma(u)$, where u is the speed of the planet relative to observer P expressed in his units. It is important to see that this choice for the power of Q in the kinetic scaling of g is consistent with Schiff's assumption[78] that the "local" speed of light is always equal to c. For light, the value of $\gamma(u = c)$ is infinite, so the result of applying the Q^{-2} scaling in eq. (143) is that the acceleration due to gravity as measured by P *is exactly zero*. This means of course that the velocity of light is indeed constant for him, as required in Schiff's method, and therefore always moves in straight line.[83]

There is another subtle point to be considered, however. The conversion factor Q is generally only equal to $\gamma(u)$ if M was previously accelerated from the rest frame of O*, *i.e.* the ORS,[47] as has been assumed above. The justification for this choice in the present application is clearly that the planet is in the gravitational field of the sun which therefore serves as the ORS in this case.

The *gravitational* scaling of g remains to be discussed. If it is assumed that g scales in the same manner as radial velocity divided by elapsed time (and therefore as S^3, see Table 2), the result obtained for the precession angle is quite different than that inferred from both experiment and GTR.[76] Consideration of Newton's definition of g in eq. (143) suggests another possibility, however. If one assumes instead that the distance from the sun in this expression scales in the same manner as L_r* in eq. (137), then a contribution of S^{-2} to the scaling factor of g is indicated. Moreover, an additional factor of S^{-1} is required if one assumes that the gravitational mass in eq. (143) scales as the inertial mass for this purpose, so that the actual gravitational scaling factor for g is S^{-3}, as indicated in Table 2. Calculation of the planetary orbits employing a standard numerical time-step procedure[79] obtains results that are in very good agreement with those observed experimentally when the scaling factors of Table 2 are employed to account for the distinctions in the units of time, length and mass of various observers in different states of motion and locations in a gravitational field. In particular, the same dependence of the precession angle of the perihelion of orbiting planets on both the average distance from the gravitational source and its mass M_S is found as in GTR.[76,79] More details of these calculations will be discussed in Chapter 11.

E. Applying the VT with Gravitational Scaling

The VT in eq. (29) has been derived in Chapter 3. C under the condition that the object and both observers in the two rest frames S and S′ move continuously at the same gravitational potential. In order to apply the VT when this condition is not met, it is necessary to use gravitational scaling in a consistent manner. Consider the case where the difference in two velocities $\mathbf{v}_3 = \mathbf{v}_2 - \mathbf{v}_1$ is to be obtained. In principle, $\mathbf{v}_2$ and $\mathbf{v}_1$ can be treated on an equal footing, but the form of the VT demands that a clear distinction be made between the velocity of the object ($\mathbf{v}_2$) and that of the second observer ($\mathbf{v}_1$). This is because it is assumed in eq. (29) that the latter is moving along the x-axis relative to the primary observer. The first step in applying the VT is therefore to rotate the coordinate system so that the latter condition is fulfilled.

When the object and the second observer are at different gravitational potentials than the primary observer, it is necessary to compute the gravitational conversion factors for the units employed in these rest frames relative to those employed by the primary observer. Assume that the corresponding value for the object is S_2 and that for the second observer is S_1. In order to obtain these values it is necessary to compute the value of A_i in eq. (130) for all three entities: A_0 for the primary observer, A_1 for the second observer and A_2 for the object. Then $S_1 = A_0/A_1$ and $S_2 = A_0/A_2$.

The VT can only be applied if the velocity of the object is expressed in the units of an observer located at the same gravitational potential (so that $v_2 = c$ if the object is a light pulse). Moreover, the coordinate system employed must satisfy the above condition regarding the placement of the x-axis. In addition, the velocity of the second observer must also be expressed in the same units as for the object (note that the final difference of velocities is to be expressed in the units of the second observer). All three of these conditions are met as follows. Let R be the three-dimensional rotation that causes the velocity of the second observer to be directed along the x-axis. The rotated velocities are then $R\mathbf{v}_2$ and $R\mathbf{v}_1$, respectively. The corresponding values expressed in the local units of the object are thus $S_2^{-1}R\mathbf{v}_2$ and $S_2^{-1}R\mathbf{v}_1$. Substituting these velocities in the VT to obtain their difference then gives the resultant velocity in the rotated coordinate system: $S_2^{-1}R\mathbf{v}_3$. The final step is to back-rotate to the original coordinate system and change to the units of the second observer, *i.e.* by forming $S_2S_1^{-1}R^{-1}(S_2^{-1}R\mathbf{v}_3)$. This result is the velocity of the object relative to the second observer expressed in his units and in the original coordinate system, as was to be obtained from application of the VT.

Chapter 11

Relativistic Calculation of the Precession of Planetary Orbits

The Universal Gravitation Theory of Newton stood unchallenged for well over a century and led to many spectacular advances in the understanding of astrophysical phenomena. In trying to improve upon Newton's theory, three main applications were considered: the gravitational red shift, the displacement of star images by massive objects, and the precession of the perihelion of Mercury's orbit around the sun. Early attempts to adapt Newton's inverse-square law (ISL) to the requirements of STR were only partially successful, as for example in Sommerfeld's treatment of planetary motion.[96,97] It has been said[98] that GTR grew out of the failure to find a suitable means of combining Newton's ISL with STR. In its final form GTR has been described as one of the greatest achievements of the human mind,[99] and the elegance of its mathematical formulation has inspired generations of theoretical physicists. Nonetheless, over the years there have been a number of attempts to find alternative gravitational theories, but subsequent experimental tests have invariably come down on the side of GTR.[100] Indeed, the very success of GTR has led to the view, held by Einstein himself,[101] that STR is a flawed theory, and that the ISL is only a good approximation whose validity does not extend beyond weak gravitational fields.

Yet both the latter theories have enjoyed remarkable success in a variety of applications, which at least raises the theoretical possibility that a viable synthesis may indeed exist. In Chapter 9 it has been shown that a simple method reported by Schiff[78] that is based solely on the EP and STR is successful in obtaining accurate results for both the gravitational red shift and the angle of displacement of light rays passing close to the sun. This method has been extended in Chapter 10 by making several substantial changes to Schiff's method (Table 2). Its application to the third key experiment, the precession of planetary orbits, will be discussed in the following.

A. Computational Procedure

The following method has been employed to compute the trajectories of objects moving in a gravitational field based on the results given in Sects. IX-X. It is assumed that the initial velocity $\mathbf{u}_o$ and position **P** of the object are known relative to a primary (stationary) observer O* located at infinity [$A_o = 1$ in eq. (130)]. A coordinate system is adopted with the sun (in the general case, the active gravitational source) located at the origin. It is further assumed that O* is stationary in the sun's rest frame. The value of the scaling quantity A_p is calculated according to eq. (130), from which the key ratio $S = A_o/A_p = A_p^{-1}$ is obtained. This allows O* to compute the *local* velocity $\mathbf{u}_p$ measured by another observer (P) who is also stationary in the sun's rest frame, but is located at the same gravitational potential as the object (planet). Based on the discussion in the Chapter 9. A, O* must take into account the difference in clock rates for the two observers. Since P's clock runs $A_p/A_o = A_p$ times *slower* than O*'s (see Table 2), his value for the object's velocity is A_p times greater ($A_p > 1$), *i.e.*, $\mathbf{u}_p = A_p\mathbf{u}_o$. This conversion is only made to obtain an initial value for $\mathbf{u}_p$. In succeeding time cycles, the value of $\mathbf{u}_p$ obtained at the end of the previous cycle will be used for this purpose.

The next step is to compute the acceleration exerted on the object by the gravitational field of the active mass. To this end it is assumed that the ISL is valid for an observer M who is at the same gravitational potential and is stationary in the object's rest frame. The corresponding value used by O* is given by eq. (143), *i.e.* by using the current value of $\mathbf{u}_p$ in conjunction with the ISL value at the location of the object, g (M).

The above information allows one to compute the change of velocity of the object over a small time interval Δt in O*'s system of standard units. To do this, however, he must use Schiff's procedure to convert $\mathbf{u}_p$ to the corresponding value in his units ($\mathbf{u}_o$), as indicated in Table 2. This means he must first resolve $\mathbf{u}_p$ into its transverse and radial components, u_p^t and u_p^r, and then *divide* these values by A_p and A_p^2, respectively. It should be noted that this is *not* just the inverse of the scaling procedure used above to obtain the initial value of $\mathbf{u}_p$ from $\mathbf{u}_o$, in which case we would simply divide all components uniformly by A_p. The reason for making this distinction will be discussed below, but first let us compute the change in the object's velocity from O*'s perspective as:

$$\Delta \mathbf{u}_o = \mathbf{g}\,(O^*)\ \Delta t\,(O^*), \tag{144}$$

with $\mathbf{g}(O^*)$ radial to the gravitational field. The velocity at the end of the time interval is then obtained by vector addition employing the VT of eq. (29). This is an important point since use of simple vector addition of $\Delta\mathbf{u}_o$ to the original value of $\mathbf{u}_o$ in each time cycle causes significant accumulation of error over a complete orbital period. In the notation introduced in Chapter 10. E, $\mathbf{u}_o$ is treated as the velocity of the "second observer" in applying the VT in this case.

The final velocity $\mathbf{u}_o'$ is then scaled using Schiff's procedure (Table 2) to obtain the corresponding local value $\mathbf{u}_p'$, that is, by multiplying the radial component by A_p^2 and the transverse by A_p. The distance Δs_o travelled by the object in the current time cycle from O*'s perspective is computed by multiplying the average velocity $\mathbf{u}_o^a = (\mathbf{u}_o + \mathbf{u}_o')/2$ by $\Delta t(O^*)$. The direction taken is that of the average *local* velocity $\mathbf{u}_p^a = (\mathbf{u}_p + \mathbf{u}_p')/2$, however, not that of $\mathbf{u}_o^a$. Note that since there is no gravitational acceleration of light in Schiff's method for computing the angular displacement of star images,[78] the magnitude of $\mathbf{u}_p^a$ is always equal to c in this case and its direction is constant as the light passes by the sun. Taking the direction the light follows to be the same as that of $\mathbf{u}_o^a$ in that application leads to inaccuracies in both the trajectory and the displacement angle, as discussed in Chapter 9.B (see Fig. 5). The final location of the object $\mathbf{P}'$ at the end of the cycle is thus computed as

$$\mathbf{P}' = \mathbf{P} + (\mathbf{u}_p^a/u_p^a)\ \Delta s_o. \tag{145}$$

It is important to see that all observers who are co-moving with O* must measure exactly the same value for $\mathbf{P}'$ according to Table

2 and the GIS. They will only disagree on the amount of elapsed time for this portion of the object's trajectory because their respective clocks run at different rates depending on their position in the gravitational field [$\Delta t(P) = \Delta t(O^*)A_p^{-1}$]. In essence, O*'s location at infinity makes him the ideal neutral observer. He and he alone can apply Schiff's scaling procedure to obtain the object's trajectory in his system of units ($A_o = 1$), and this information can then be converted to the units of any other observer simply by knowing the latter's value of A_p.

In the specific computational approach adopted in the present work, there is another matter that needs to be clarified, however. Both $\mathbf{u}_o$ and $\mathbf{u}_p$ are continuous functions of time, but only one of them can remain the same in going from the end of one time cycle to the beginning of the next. This is because the distance of the object from the active mass is constantly changing, and therefore the value of the scaling parameter A_p generally varies between successive cycles. In view of the success of Schiff's approach to the calculation of the displacement of star images caused by the gravitational field of the sun, in which case both the magnitude (c) and the direction of the local light velocity $\mathbf{u}_p$ are held constant throughout, it seems preferable at the beginning of each cycle to set $\mathbf{u}_p$ equal to the value of $\mathbf{u}_p'$ at the end of the previous one, as already mentioned. In so doing, one must accept the fact that this choice generally precludes the existence of a similar equality between the corresponding values of $\mathbf{u}_o$ and $\mathbf{u}_o'$ for the primary observer in going from one time cycle to the next, but this is inconsequential because in the last analysis these quantities as defined are only an artifact of Schiff's method.

The above description also provides some rationale for scaling the radial component of the velocity $\mathbf{u}_p$ differently than the transverse. When the planet moves radial to the gravitational field, the value of A_p is changing continuously, whereas by definition, transverse motion does not change r and thus leaves A_p constant. Therefore the normal scaling of velocity according to eq. (135) and Table 2 (n = 1 for velocity) is required for the transverse component. The extra factor of A_p for the radial component is an *ad hoc* adjustment that has proven to accurately account for the variation of this quantity with distance from the active mass. Instead of allowing the value of A_p to vary during a time cycle, it proves sufficiently accurate to a) scale u_p^r with A_p^{-2} instead of A_p^{-1}

and b) use this result *only* to obtain u_o^r for observer O* located in a gravity-free region of space.

B. Results of the Calculations

The above procedure has been applied to the calculation of the relativistic contribution to the precession angle of the perihelion of planetary orbits around the sun. At the start of the calculation the position and velocity of the planet are taken from experiment (based on the observed values for the mean radius r and eccentricity e of a given orbit). The solar mass is taken to be 1.99×10^{30} kg, whereas the mass of the planet is not required, consistent with the unicity principle. The time interval Δt(O*) for each cycle in the numerical procedure has been varied in all cases to insure that a proper degree of convergence is obtained for the calculated results (quadruple precision has been used in all computations).

The value of the precession angle Θ of the perihelion of Mercury's orbit around the sun obtained in the present treatment is 43″.0033/cy, in good agreement with both the currently accepted experimental value for this quantity of 43″.2 ± 0″.9/cy[102] and that computed by Einstein from GTR of 43″.0076/cy.[76,103] In the latter work he obtained a closed expression[103,104] which indicates that the precession angle in general is proportional to M_s and inversely proportional to both r and $(1 - e^2)$. Tests have therefore been carried out for different values of the latter three quantities and very good agreement with the predictions of GTR has been found in all cases. Indeed, since the amount of computer time required increases with r, most of the tests carried out are for a hypothetical planet with one-thousandth of Mercury's radius and therefore a period of revolution around the sun of only 240 s. When the solar mass is increased by a factor of 10.0, it is found that the value of Θ is 10.0012 times greater. If the mean radius is cut in half, Θ is found to increase by a factor of 1.9990. Similarly good agreement with GTR is obtained if the radius is changed by factors of 10 and 100. Finally, when the eccentricity e of the orbit is changed from its experimental value of 0.2056 for Mercury to 0.10, the value of Θ is found to be 0.9677 times smaller, as compared to the predicted factor of 0.9674.

The A_p factors have been computed in the present treatment in two different ways: by means of eq. (130) in each time-step, or by making use of the proportionality relationship of eq. (132) after using eq. (130) to obtain an initial value only. The corresponding two values of Θ agree to within a factor of 1.000093, with that obtained with the latter definition being higher. This result thus clearly supports the conclusion that the whole concept of gravitational scaling is rooted in the conservation of energy principle (see Chapter 10. A).

One can summarize the above results as follows. The present theoretical approach obtains results that are consistent with all known measurements of perihelion precession angles, *including those of the earth and Venus*. They are also in nearly quantitative agreement with the predictions of GTR for the same quantities. The procedure employed can be viewed as a generalization of Schiff's method[78] for computing the displacement angle of star images during solar eclipses (strictly speaking, what is actually calculated via Huygens' principle is the angle by which the wave front of the light rotates relative to its starting orientation at the star[83,105]). Corresponding tests have been carried out with the present procedure for an object moving with local speed c, and excellent agreement with Schiff's (and therefore Einstein's) result has been obtained, including the dependence of this angle on the mass of the gravitational source and the distance of closest approach by the light. A more detailed discussion of these results for the displacement of star images is given in Chapter 9. B.

C. Some General Remarks about the Scaling Procedures

For nearly an entire century physicists have held steadfastly to the belief that it is impossible to construct a viable relativistic gravitational theory in the framework of Euclidean space which is based on Newton's ISL and Einstein's STR. The methodology employed in Chapter 9. A has investigated this position anew by assuming the gravitational isotropy of space (GIS), *i.e.* the principle that all observers who are not in relative motion to one another must agree on measurements of distance. A snapshot of the universe at a given time by one such observer will therefore be indistinguishable from that of any other fulfilling this condition.

Einstein's STR has been modified to guarantee remote simultaneity by replacing the LT with the ALT of eq. (62), so that the condition "at the same time" is attainable in practice.

When an observer changes his state of motion, however, the distances between fixed points in the universe do change. If clocks run slower in his new rest frame, he will find that the universe has shrunk because his unit of distance has increased proportionately. This relationship is necessary to satisfy the scaling relation for velocities in eq. (96) as well as the light-speed postulate. Consistent with the first paragraph, observers located at other gravitational potentials in the same state of motion will agree completely on the values of all distances between fixed points.

The situation with velocities is different. As an observer lowers his position in a gravitational field, he finds that velocities increase uniformly. Everything appears to speed up for him as well as for all other observers making the same transition. A change in his state of motion does not alter the velocities he measures for other objects, however. The latter results are all consistent with the way the rates of clocks change with their location in a gravitational field and/or their state of motion.

These are but two examples of the way physical properties change with the scaling factors listed in Table 2. Gravitational mass is exceptional because all observers agree on the value of this quantity for any given object. This is a key condition for using the ISL. It allows one to calculate the local value of g for a given object, which can then be converted to the system of units employed by observer O* at infinity (with $A_o = 1$) who is stationary in the rest frame of the active mass. The acceleration due to gravity in these units is thus obtained by multiplying the local value by a factor of $S^{-3} = (A_p/A_o)^3 = A_p^{\ 3}$. There is an interesting qualitative conclusion that can be drawn on this basis, namely that for strong fields a purely Newtonian, non-relativistic description of the motion can greatly *underestimate* the effects of gravity on a given object by failing to adjust the local g-field by the above factor.

The trajectory calculations provide a justification for employing a coordinate system in Euclidean space in which all objects of the universe can be located uniquely. As long as one takes proper account of the fact that the units of time, velocity and acceleration vary with one's position in a gravitational field, it is then possible to carry out trajectory calculations exclusively in Euclidean space. The necessary adaptation can be accomplished by inserting a small number of statements in a comparatively simple computer program which otherwise treats planetary motion strictly on the basis of Newton's ISL.

The development of a comprehensive gravitational theory that relies on the local validity of the ISL inevitably raises questions about whether such forces can be transmitted instantaneously across long distances. Newton himself rejected such an interpretation in the strongest terms, but this did not keep him from using the ISL to solve longstanding problems in astrophysics. The fact remains, however, that the above computer program uses time intervals as small as 10^{-4} s to calculate the change in velocity of a planet caused by the sun which is as much as 7×10^{10} m distant. It is a matter of opinion whether GTR succeeds in eliminating the need for "action at a distance" by introducing the concept of "curved space-time." The present work indicates that the units of physical quantities vary in a precisely predictable manner with the distance of a given location from the gravitational source, suggesting that something like a distance-dependent stationary field exists *at all times* and therefore does not need to be transmitted to have its effect on any object that is located at that point in space. It has demonstrated that, with proper attention to detail, it is possible to obtain a level of accuracy in trajectory calculations by appropriate scaling of the above physical units that is comparable to that of GTR. This experience speaks for the validity of the assumptions that form the basis for arriving at this synthesis, and at least underscores the practicality of the ISL that Newton so skillfully exploited during his lifetime.

D. Application to the Global Positioning System (GPS)

There are two causes of clock-rate variations in terrestrial experiments. Both of them involve the earth's center of mass (ECM), one depending exclusively on the speed of a given clock relative to that position and the other on the corresponding difference in gravitational potential.[42] As a result, in comparing different clocks on the earth's surface, it is necessary to know both the latitude χ of each clock as well as its altitude h relative to sea level.[105] The time-dilation effect is governed exclusively by the UTDL of eq. (57), whereby the speed to be inserted in the respective γ factors is equal to $R_E\Omega\cos\chi$, where Ω is the earth's rotational frequency (2π radians per 24 h = 86 400 s) and R_E is the earth's radius (or more accurately, the distance between the location of the clock and the ECM). A standard clock S needs to be designated. Theoretically, there is no restriction on its location. Its latitude χ_S (as well as its altitude r_S relative to the ECM) is then an important parameter in computing the ratio of the rates of each clock in the network with that of the standard clock. It is helpful to define the ratio Q as follows:

$$Q = \gamma(R_E\Omega\cos\chi)/\gamma(R_E\Omega\cos\chi_S). \tag{146}$$

According to eq. (57), this ratio tells us how much slower (if $Q > 1$) or faster (if $Q < 1$) the given clock runs than the standard if both are located at the same gravitational potential. The gravitational red shift needs to be taken into account to obtain the actual clock-rate ratio, however. For this purpose, it is helpful to define a second ratio S for each clock:

$$S = 1 + g(r - r_S)c^{-2}, \tag{147}$$

where r is the distance of the clock to the ECM [see eq. (129)]. This ratio tells us how much faster ($S > 1$) the clock runs relative to the standard by virtue of their difference in gravitational potential. The elapsed time Δt on the clock for a given event can then be converted to the corresponding elapsed time Δt_S on the standard clock by combining the two ratios as follows:

$$\Delta t_S = QS^{-1}\Delta t. \tag{148}$$

It is possible to obtain the above ratios without having any communication between the laboratories that house the respective clocks. The necessary synchronization can begin by sending a light signal directly from the position of the clock A lying closest to the standard clock S. The corresponding distance can be determined to as high an accuracy as possible using GPS. Division by c then gives the elapsed time for the one-way travel of the signal based on the standard clock S. The time of arrival on the standard clock is then adjusted backward by this amount to give the time of emission $T^{S0}(A)$ for the signal, again based on the standard clock. The corresponding time of the initial emission read from the local clock is also stored with the value $T^0(A)$. In principle, all subsequent timings can be determined by subtracting $T^0(A)$ from the current reading on clock A to obtain Δt to be inserted in eq. (148). The time T^S of the event on the standard clock is then computed to be:

$$T_S = QS^{-1}\,\Delta t + T^{S0}(A), \qquad (149)$$

where Q and S are the specific values of the ratios computed above for clock A.

Once the above procedure has been applied to clock A, it attains equivalent status as a standard. The next step therefore can be applied to the clock which is nearest either to clock A or clock S. In this way the network of standard clocks can be extended indefinitely across the globe. Making use of the "secondary" standard (A) naturally implies that all timings there are based on its adjusted readings. It is important to understand that no physical adjustments need to be made to the secondary clock, rather its direct readings are simply combined with the Q and S factors in eq. (149) to obtain the timing results for a hypothetical standard. A discussion of this general point has been given earlier by van Flandern.[106] The situation is entirely analogous to having a clock in one's household that runs systematically slower than the standard rate. One can nonetheless obtain accurate timings by multiplying the readings from the faulty clock by an appropriate factor and keeping track of the time that has elapsed since it was last set to the correct time. The key word in this discussion is "systematic." If the error is always of quantitatively reliable magnitude, the faulty clock can replace the standard without making any repairs.

The same principles used to standardize clock rates on the earth's surface can also be applied for adjusting GPS satellite clocks. Assume that the clock is running at the standard rate prior to launch and is perfectly synchronized with the standard clock (*i.e.* as adjusted at the local position). In order to illustrate the principles involved, the gravitational effects of other objects in the neighborhood of the satellite are neglected in the following discussion, as well as inhomogeneous characteristics of the earth's gravitational field. The main difference relative to the previous example is that the Q and S factors needed to make the adjustment from local to standard clock rate using eqs. (148,149) are no longer constant. Their computation requires a precise knowledge of the trajectory of the satellite, specifically the current value of its speed v and altitude r relative to the ECM. The acceleration due to gravity changes in flight and so the ratio S also has to be computed in a more fundamental way. For this purpose, it is helpful to define the following quantity connected with the gravitational potential, which is a generalization of the quantity in parentheses in eq. (139):

$$A\,(r) = 1 + GM_E/c^2r, \tag{150}$$

where M_E is the gravitational mass of the earth (5.975×10^{24} kg). The value of S is therefore given as the ratio of the A (r) values for the satellite and the standard clock:

$$S = A\,(r_S)/\,A\,(r), \tag{151}$$

which simplifies to eq. (147) near the earth's surface (with $g = GM_E r_S^{-2}$). The corresponding value of the Q ratio is at least simple in form:

$$Q = \gamma(v)/\gamma(v_S) = \gamma(v)/\gamma(R_E\Omega\cos\chi_S). \tag{152}$$

Note that the latitude χ_S is not that of the launch position relative to the ECM, but rather that of the original standard clock in its current position. The accuracy of the adjustment procedure depends primarily on the determination of the satellite speed v relative to the ECM at each instant.

In this application the underlying principle is to adjust the satellite clock rate to the corresponding standard value over the entire flight, including the period after orbit has been achieved.[105] The correction is made continuously in small intervals by using eq. (148) and the current values of Q and S in each step. The result is tantamount to having the standard clock running at its normal rate on the satellite. The above procedure super-cedes the "pre-correction" technique commonly discussed in the literature[106,107] according to which the satellite clock is *physically* adjusted prior to launch. The latter's goal is to approximately correct for the estimated change in clock rate expected if the satellite ultimately travels in a constant circular trajectory once it achieves orbital speed. The present theoretical procedure has the advantage of being able to account for departures from a perfectly circular orbit and also for rate changes occurring during the launch phase.

E. Test of GRT for Rotating Bodies: Challenge for GTR

Thomas spin precession[27] has been discussed in Chapter 3. D. Although the derivation of this effect has focused on the transverse acceleration experienced by an orbiting electron, it also has application to the rotation of macroscopic bodies.[26] At about the same time he introduced his method for calculating the angle of deflection for light in a gravitational field,[78] Schiff[29] proposed an interesting experiment to investigate this possibility. It is based on eq. (41) for the angular velocity ω_T of a spinning electron. According to Newton's ISL, the acceleration **a** of an orbiting satellite is given as:

$$\mathbf{a} = -\,GM_E\mathbf{r}/r^3, \tag{153}$$

where **r** is the distance vector to the ECM. Substitution of this value in eq. (41) gives, with $\gamma \approx 1$ because of the relatively low orbital speed of the satellite,

$$\omega_T = (GM_E/2c^2r^3)\ \mathbf{v} \times \mathbf{r}, \tag{154}$$

for the precession rate in the plane of its orbit. Schiff points out[29] that the corresponding result based on GTR is to a good approximation:

$$\omega_T = (-3GM_E/2c^2r^3)\ \mathbf{v} \times \mathbf{r}. \tag{155}$$

Comparison shows that the GTR value for the precession frequency is three times larger and, even more interestingly, in the opposite sense as that calculated using the ISL to compute **a**. The expression for the precession frequency in eq. (154) is derived from STR by performing successive Lorentz transformations, whereby the result is the same for the ALT as for the LT since it holds for any choice of ε.[28] It seems very strange that the sense would be opposite for a satellite as for an electron. After all, they differ primarily in the values of their respective gravitational masses. Since Thomas precession of an electron is well established experimentally, this raises the question of why the same relativistic formula for the precession frequency in eq. (154) cannot be applied to macroscopic bodies.

The extension of Schiff's method for calculating the precession of the orbits of planets has been shown to agree quantitatively with the results of experiment in Chapter 11. B, and there is no reason to expect that the sense of the rotation would depend on the mass of the orbiting object according to this theory. Thus far in discussing the results of the extended Schiff methodology, the conclusion has always been that they are in quantitative agreement with those of GTR. Since that is not the case in the present application, a clear opportunity presents itself for a definitive experimental test to decide which theory is more accurate.

F. Extension to Treatment of Many Bodies: Third Law

The computational treatment discussed in Chapter 11. A only deals with a single passive mass (Mercury or a light wave) and single active mass (the sun). The steps that must be taken to extend the method to more than one planet will now be considered. At the outset, it should be clear that there is a fundamental error that needs to be eliminated in the original procedure, namely the very concept that one can look upon the planet as being in a qualitatively different class than the sun. While this is a very good approximation in the present application, it fails to pass the requirements of Newton's Third Law. The sun is also affected by the gravitational field of its planets and therefore should only be distinguished from the latter by virtue of its physical properties, specifically its overwhelmingly greater mass. In short, the computational method should treat the sun and the planets on an equal footing in principle, thus including the gravitational interaction in which the sun is the passive mass and the planet is the active mass.

The above requirement can actually be satisfied in a relatively straightforward manner in Schiff's model,[78] which is therefore a distinct advantage of the accompanying theory as a whole. All that is necessary is to carry out the calculation for the sun being acted upon by its planet(s) in a completely analogous manner in each time cycle as in the original treatment of the precession of the planetary orbit of Mercury described in Chapter 11. A. Indeed, there is no conceptual difficulty in making this extension to an arbitrary number of bodies.

In terms of a functioning computer program, one has to simply add a do loop to the code which runs from 1 to the number of participating objects n. The instructions inside the loop must be applicable to each object in a completely equivalent way. The passive object's location vector in the three-dimensional coordinate system and corresponding local velocity are all that is required in each case (a total of six values). The gravitational masses of all participating objects are also made available at the start of the calculation. The local value of **g** for each passive mass is obtained using the ISL by considering all other participating objects (active masses), *i.e.* excluding the object under current consideration (no self-interaction). First, the individual $\mathbf{g}_i$ values for each active mass V_i must be computed separately, *i.e.* by only considering the relationship between V_i and the passive mass. The total value of **g** is then computed by summing over all the above $\mathbf{g}_i$ values using ordinary vector addition.

The location (x,y,z) of the origin of the total **g** field must next be determined. This is done by computing the weighted average $x^a = g^{-1}\Sigma g_i x_i$ for each of the three coordinates. The resulting vector $\mathbf{r}^a = (x^a, y^a, z^a)$ then plays the role of the sun's location in the computational procedure described above for the calculation of Mercury's orbit. In summary, it is assumed that a gravitational field **g** is located at position $\mathbf{r}^a$ for the purposes of computing the interactions with the current passive mass.

There is still one open point that requires discussion before moving on to the trajectory calculation, however. According to Table 2, it is necessary to know the value of the kinetic scale factor Q in order to determine g (O*) in eq. (144), whereby $Q = \gamma(\mathbf{u}_p)$ in the Mercury orbit application. The UTDL of eq. (57) is used for this purpose, but it is first necessary to assign an ORS for the calculation of the v_{i0} speeds in the pertinent γ factors in this relation. In general, the ORS will be moving in the assumed coordinate system, so its velocity $\mathbf{v}_{ORS}$ must be known. In addition, a velocity $\mathbf{v}^a = (v_x{}^a, v_y{}^a, v_z{}^a)$ must be determined for an object that is formally responsible for the total **g** field located at $\mathbf{r}^a$. In the original procedure $\mathbf{v}^a = 0$ is assumed for the velocity of the sun, *i.e.* the sun is assumed to be stationary at the origin of the coordinate system. In the present (general) case, one can only make a plausible assumption to obtain $\mathbf{v}^a$, such as carrying out a g-averaged procedure analogous to that used to determine $\mathbf{r}^a$. Then the v_{i0} relative speeds to the ORS must be calculated using the VT for both $\mathbf{v}^a$ and $\mathbf{u}_p$ for the passive mass. The resulting v_{i0} quantities are then substituted into eq. (57) to obtain the ratio Q of γ factors. Note the following convention: if the passive mass is moving faster relative to the ORS, as is the case in the Mercury orbit calculation, then $Q > 1$; in the opposite case, $Q < 1$.

The gravitational scaling procedure requires a clear distinction between the radial and transverse velocity components (see Table 2) of the passive mass. The above value of **g** is used to make this definition unique in each case, *i.e.* the radial direction is defined to be parallel to **g**. This information is also sufficient for computing the local A_p value. For this purpose, it is helpful to define GM_S/r_O in eq. (130) as gr, where r is the distance between the location of the passive mass and $\mathbf{r}^a$. The computational procedure described in Chapter 11. A is then applied for each passive mass in succession. The final values of the vector location **P′** and the local velocity $\mathbf{u}_p{}^a$ for each object at the end of every time cycle are then computed [see eq. (145) and the discussion immediately preceding it].

There is still one aspect in the above generalized procedure that needs to be discussed, namely where to place the origin of the three-dimensional coordinate system. Ideally, one would like the details of the objects' trajectories that result from the calculations to be completely independent of this choice. Analysis of the above procedure shows *that only distances between locations* are ever used, specifically in the critical determinations of the relevant A_p and **g** values for each object (note that velocities are not affected by a change in coordinate system). As a result, there is no question that the procedure satisfies the condition of coordinate-system invariance. This is clearly accomplished by its use of the weighted average technique for each set of active masses to make the above determinations, rather than trying to come up with an alternative procedure that treats each active mass individually. There has been much discussion in the literature in general[108] as how to design a generally applicable gravitational theory that is both suitably relativistic and coordinate-independent, so this aspect of the present computational procedure should not be underestimated.

Chapter 12

Uniform Scaling of the Properties of Light

The story of how Einstein first came to the conclusion that translational acceleration is indistinguishable from a uniform gravitational field has been repeated many times.[94,109,110] He introduced the concept[38] by considering two different coordinate frames, one of which is at rest in a "homogeneous gravitational field" while the other is subject to a constant acceleration. Since his paper in 1907, generations of students have gone through the mental gymnastics employed by Einstein that are required to properly understand the basis of what has come to be known as the Equivalence Principle (EP). His arguments were based in large part on his earlier work on STR,[5] as well as on Galileo's unicity principle and Newton's gravitational theory.

Einstein consistently used the properties of light as a guide in developing the special and general theories of relativity,[109,111] particularly its speed in free space. The first application of the EP[38] was the calculation of the gravitational red shift. In this case, three different rest frames are required to complete the argument. As Pais has pointed out,[94] two of these frames are inertial, so it was possible to apply STR directly to them, while measurements in the third could be identified with one of the latter during a small time interval. He was able to predict on this basis that the rates of clocks increase in a well-defined manner as they are brought to a higher gravitational potential [see eq. (139)], and this assessment was verified by the Pound-Snider experiment[91] discussed in Chapter 10. C. He pointed out that "generators of spectral lines" serve as practical examples of such clocks that are readily available at different gravitational potentials, and he therefore concluded that light at the solar surface has a "longer wavelength than that generated terrestrially" from the same source on earth. He then went on in the following section[38] to prove that the speed of light increases by the same factor as the light frequency when the source is moved to a higher gravitational potential. For this purpose he used the transformation properties of the Maxwell equations and again applied the EP to reach his conclusion regarding the effect of gravity on light speed.

The gravitational scaling procedure draws heavily on Einstein's arguments based on the EP. The conversion factors listed in Table 2 are usually deduced in one way or another from application of the EP to the properties of light. This procedure does not always lead to the correct results, however, as will be discussed in the following section.

A. Transverse Doppler Experiments Employing Rotors

The experiments carried out by Hay *et al.*[37] and analogous work by Kündig[39] and Champeney *et al.*[40] have been discussed in Chapter 5. A. The empirical formula obtained for the frequency shift in their transverse Doppler effect (TDE) study employing high-speed rotors is given in eq. (49). Since the absorber was located at the rim of the rotor ($R_a > R_s$), it follows that a frequency shift to the blue was measured, *i.e.* $\Delta\nu/\nu > 0$. It has already been mentioned that this result stands in contradiction to the red shift expected from STR on the basis of eq. (26), as pointed out by Sherwin,[41] although this state of affairs was ignored in the original experimental reports.[37,39,40]

Instead, the observed blue shift was explained on the basis of the EP and the gravitational red shift. Accordingly, the centrifugal force experienced by the detector corresponds[111] to a gravitational field of strength $\varrho\omega^2 = v^2\varrho^{-1}$ (ω is the rotor frequency, ϱ is the distance of the absorber from the rotor axis and $v = \varrho\omega$). Integration between the source at ϱ_X and the absorber at ϱ_D leads to the following relations:

$$\Delta\nu/\nu = -\Delta\phi c^{-2} = 0.5c^{-2}(v_D{}^2 - v_X{}^2), \qquad (156)$$

based on the relation between frequencies and potential difference Φ in eq. (139). The fractional frequency shift therefore has the same magnitude as in the STR formula of eq. (26) as long as $v_X = 0$, but it has the opposite sign (blue, not red). In the rotor TDE experiments, the absorber/detector moves faster than the x-ray source and thus the positive value observed for $\Delta\nu/\nu$ is indeed expected on the basis of the EP and eq. (156).

It is useful to make the same analysis of the measurement process for different observers in the above experiment as has been made in Chapter 10. C for the Pound-Snider counterpart.[91] The observers at the x-ray source and detector will again be referred to as X and D, respectively. The two events of interest are the emission of the x-rays (III) and their subsequent detection (IV), analogous to events I and II in Chapter 10. C for the Pound-Snider experiment. The values measured by X at the initial emission process are thus: ν_X (III) = ν, λ_X (III) = λ and c_X (III) = c, where, as in Chapter 10. C, ν and λ are the standard frequency and wavelength of the light source. The question is: what does the same observer measure for event IV? There can be no change in frequency for the same reason as Einstein gave in his 1907 article.[38] At the same time, his second postulate of STR[5] demands that the speed of light also be unchanged. Since it must further be assumed that the speed of light waves in free space is always equal to the product of their frequency and wavelength, the conclusion is that nothing changes for observer X as the x-rays move between the source and absorber, *i.e.* ν_X (IV) = ν, λ_X (IV) = λ and c_X (IV) = c.

The clock at the detector in the rotor experiment runs at a slower rate than at the light source,[39] however. There is no ambiguity about this, as Sherwin has pointed out.[41] As a consequence, one can proceed as before with the Pound-Snider experiment and assign different units of time to the two observers. According to eq. (156), the conversion factor in the TDE case is $S = 1 + 0.5c^{-2}(v_D^2 - v_X^2)$. If $v_X = 0$, it has the same numerical value as for the gravitational red shift experiment discussed first. Observer D therefore is expected on this basis to measure the frequency for both events to be S times larger than observer X. But what about the light speed? Even though the detector is subjected to a high degree of acceleration, there is no reason to think that Einstein's second postulate of STR does not continue to hold for him[45]. This forces one to conclude that the corresponding wavelength values must be smaller by the same factor S for each event. As a result,

Table 3. Comparison of measured frequency, wavelength and light speed for two observers X (located at the x-ray source) and D (located at the Mössbauer absorber) for four different events: I (emission) and II (absorption of x-rays in the Pound-Snider gravitation experiment[91]), and III (emission) and IV (absorption of x-rays in the Hay *et al.* rotor experiment[37]). The scaling parameter S is equal to $1 + \Phi c^{-2}$ for events I and II [see eqs. (139,140)], and has the same value for events III and IV because of the EP relation.

Observer	Event	Frequency	Wavelength	Light Speed
X	I	ν	λ	c
X	II	ν	λS^{-1}	cS^{-1}
X	III	ν	λ	c
X	IV	ν	λ	c
D	I	$S\nu$	λ	Sc
D	II	$S\nu$	λS^{-1}	c
D	III	$S\nu$	λS^{-1}	c
D	IV	$S\nu$	λS^{-1}	c

one comes to the following conclusions for the measured values of observer D: ν_D (III) = ν_D (IV) = $S\nu$, λ_X (III) = λ_X (IV) = $S^{-1}\lambda$, c_X (III) = c_X (IV) = c.

The results for both sets of measurements are summarized in Table 3. If the EP were to hold strictly, it would be possible to simply interchange the measured values for events I and III for any given observer. The same would hold true for events II and IV. In fact, this is only true for events II and IV in the case of observer D, and for events I and III for observer X. The latter relationship simply underscores the fact that *in situ* measured values do not change with either the state of motion of the observer or his position in a gravitational field (locality principle). The reason that the EP does not hold for the other two pairs of cases (events I and III for D and events II and IV for X) is clearly because the speed of light is not affected in any way by centrifugal acceleration, whereas it does depend on the location of the source in a gravitational field. The EP does hold for frequencies in all four cases, both for X and D, and this has been the point of emphasis, both in the original TDE study employing a rotor[37] and in subsequent theoretical discussions of this experiment such as Sherwin's[41] and Sard's.[111] However, the situation with wavelengths is far less sanguine, since different results for this quantity are obtained in the above two comparisons for which the light speed is not the same.

The above comparisons indicate that the *strict* interpretation of the EP that is incorporated in the phrase "Einstein's elevator",[75,94] *i.e.* that all physical phenomena in two rest frames are *completely* equivalent in which there is, respectively, a uniform gravitational field **g** in one, and no gravitational field but a translational acceleration **–g** in the other, *is incorrect*. However, the TDE experiments with rotors show that gravity and acceleration have something else in common that can be exploited to good advantage. They both produce changes in clock rates and other physical properties that are completely unambiguous to all observers. One knows, for example, that the accelerated clock in the rotor experiment[38] runs slower than when it is at rest in the laboratory and that the fractional change can be determined quantitatively by means of eq. (156), in complete analogy to eq. (139) for the case when clocks change their position in a gravitational field[91] (note that the ambiguity in clock rates foreseen in STR precludes the use of rational units and the PRM of Chapter 4. A in describing relationships between measured values). By extension, one can expect that the units of other physical properties vary with acceleration in a well-defined manner, *i.e.* the concept of *uniform scaling* applies equally well in this case as for gravitational effects, as already has been concluded in Chapter 7. C.

To be more specific, one must first define a quantity Q similar to S in Table 3 that applies to "kinetic" variations in properties induced by acceleration (see Table 2). According to eq. (156), the obvious candidate for the rotor experiment is:

$$Q = 1 + 0.5c^{-2}(v_D^{\;2} - v_X^{\;2}), \qquad (157)$$

which is also consistent with the UTDL of eq. (57). Accordingly, S should be replaced by Q in the entries in Table 3 for cases D III and D IV. In summary, the rotor experiments are seen to stand in contradiction to the EP and Einstein's elevator, while at the same time verifying the uniform scaling procedure in the ALT version of relativity theory.

Table 4. Dependence of various properties of light on the positions of the object and observer in a gravitational field and on the corresponding clock rate parameters α_O and α_M (see text for definitions). The corresponding dependence for *in situ* measurements ($A_O = A_P$ and $\alpha_M = \alpha_O$) is given in the third column, whereas in the fourth, the corresponding results are given for the case of light in free fall in a gravitational field from point P to P' for the same (stationary) observer ($A_{P'} > A_P$). The powers of S and Q are given in parentheses below each quantity (see Table 2 for definitions).

Quantity	General Value (not in free Fall)	*In Situ* Value	Free Fall Value (emitted from P)
Speed of Light (1, 0)	$(A_O/A_P)c$	c	$(A_O/A_{P'})c$
Wavelength (0, 1)	$(\alpha_M/\alpha_O)\lambda$	λ	$(\alpha_M/\alpha_O)(A_P/A_{P'})\lambda$
Frequency (1, –1)	$(\alpha_O/\alpha_M)(A_O/A_P)\,v$	v	$(\alpha_O/\alpha_M)(A_O/A_P)v$
Energy (1, 1) ("E = h v")	$(\alpha_M/\alpha_O)(A_O/A_P)E$	E	$(\alpha_M/\alpha_O)(A_O/A_P)E$
Momentum (0, 1) ("p = h/λ")	$(\alpha_M/\alpha_O)\,p$	p	$(\alpha_M/\alpha_O)(A_{P'}/A_P)p$
Planck's Constant (0,2)	$(\alpha_M/\alpha_O)^2h$	h	$(\alpha_M/\alpha_O)^2h$
Accel. Due to gravity (–3, –2)	$(\alpha_O/\alpha_M)^2(A_P/A_O)^3g$	g	g = 0 for light

B. Overview of Scaling Results

Having determined the manner in which the units of various physical quantities vary with changes in the state of motion of the observer and his location in a gravitational field, it is of particular interest to consider what influence these conclusions have for light pulses/photons. A summary of these results is given in Table 4. It is assumed that the light source is stationary in the rest frame of inertial system M, with kinetic scaling factor $\alpha_M = \gamma(v_{M0})$ [the subscript 0 refers to the ORS in eq. (57)], and is located at point P in the gravitational field, whereas the observer is located at point O in an inertial system with kinetic scaling factor $\alpha_O = \gamma(v_{O0})$ [$Q = \alpha_M/\alpha_O$].

The speed of light observed by O is proportional to the ratio $S = A_O/A_P$ (see Table 2). In accordance with Einstein's second postulate of STR,[5] when the source and the observer are at the same gravitational potential ($A_O = A_P$), the value of the speed of light is always equal to c, regardless of the magnitude of their relative speed. The different rates of clocks in the respective rest frames of the source and the observer that are caused exclusively by the time-dilation effect have no influence on the measured value, consistent with the results of the Michelson-Morley and Kennedy-Thorndike experiments.[3,51] This is the justification for concluding in Table 1 that lengths scale as Q, the same ratio as for elapsed times. The dependence of the speed of light on location in a gravitational field given above and in Table 2 is the same as deduced by Einstein in his 1907 and 1911 papers.[38,75] It is greater than c when the source is located at a higher potential than the observer O, and it is smaller when it is located at a lower potential as in the classic example of light passing close to the sun.

The wavelength of light is proportional to $Q = \alpha_M/\alpha_O$, *i.e.* it is dependent on both the state of motion of the source and the observer, as shown in Table 4. The *in situ* value for a given source is the same for all observers, however, consistent with the RP and experiment.[3] The transverse Doppler effect[34,35,36] allows a verification for the dependence on Q. There is an additional factor of $(1 \pm u/c)$ if the light source is moving directly away/toward the observer O due to the motion of the latter into the waves (first-order Doppler effect). The wavelength of light is independent of the location of both the source and the observer in a gravitational field. As discussed in Chapter 6. D, the Doppler factor mentioned above is not actually a relativistic effect, but simply reflects the fact that the light source is moving either out of or into the wave pattern.

The variation of the light frequency must be such as to satisfy the rule that in free space it is always equal to the quotient of the light speed and the corresponding wavelength. It is, therefore, proportional to both $Q^{-1} = \alpha_O/\alpha_M$ and $S = A_O/A_P$ (see Table 4). There is also a first-order Doppler factor in this case if the light source is moving toward or away from O. It is the reciprocal of that required for wavelengths, consistent with the light-speed constancy requirement.

In order to deduce the variation of the energy and momentum of the photons it is necessary to know how Planck's constant varies with the state of motion of the source and with its relative location in the gravitational field. Since energy is a scalar quantity, its value must be independent of the direction in which the light travels, whereas the light frequency is subject to the Doppler effect. This means that the value of Planck's constant must vary with the orientation of the light velocity to the observer as well. A simple means of circumventing this complication is to restrict one's attention to the case of the transverse Doppler effect. This eliminates the $(1 \pm u/c)$ factor from the calculations. Further, only light moving perpendicular to the gravitational field needs to be considered to establish the value of the required energy/frequency ratio, from which it can be concluded that its value is proportional to Q^2, as already discussed in Chapter 7. C [see (eq. 102)]. This relationship can be tested experimentally by means of the photoelectric effect. The variation of the energy of the photons is then comparatively simple. It is proportional to both Q and S, independent of the orientation of the light velocity to both the observer and the gravitational field.

The magnitude of the corresponding momentum of the photons can be obtained by using the same value for Planck's constant in the relation "p" = "h/λ." The quotation marks are used to indicate that the value to be inserted for each quantity is that actually measured by the observer in a given situation. On this basis, "p" is found to be proportional to $Q = \alpha_M/\alpha_O$ and is completely independent of the location of both the light source and the observer relative to the gravitational source (momentum scales in the same manner as length; see Table 2). Note that there is no Doppler effect for either momentum or energy, *i.e.*, no $(1 \pm u/c)$ factor is required in either case.

The corresponding values for the above quantities that are observed *in situ* are shown in the third column of Table 4. These values are obtained from the previous (general results) by setting $A_O = A_P$ and recalling that the observations are being made by observer M in each case (*i.e.*, $\alpha_M = \alpha_O$). In accord with the RP, they are all independent of both the state of motion of the observer and his location in a gravitational field.

On the basis of the results discussed above, it is possible to make definitive predictions regarding the variation of the properties of light when the source is lowered (as opposed to dropping in free fall) in the gravitational field (without altering its velocity). Let us assume that the source is initially at point P in the gravitational field and is then lowered to point P′[$A_{P'} > A_P$, see eq. (130)].

Table 5. Dependence of various properties of light on the positions of the source and observer in a gravitational field and their relative states of motion on the kinetic scaling factor $Q = a_M/a_O$ and the gravitational scaling factor $S = A_O/A_P$ (see Table 2 and text for definitions). The first two rows contain results for the case when the light source (or other object) begins free fall at point P ($S > 1$) in the gravitational field. The succeeding two rows give the corresponding results when the light source is in free fall and has reached point P' in the field [$\gamma = A_{P'}/A_P > 1$, see Table 4 and eq. (132)].

Observer	Light speed/ Velocity	Energy	Freq.	Momentum	Wavelength	Inertial Mass	Planck's Constant
In situ P	c	E	ν	p	λ	m_I	h
O	Sc	QSE	$S\nu/Q$	Qp	$Q\lambda$	Qm_I/S	Q^2h
In situ P'	c	E	ν	p	λ	m_I	h
O	Sc/γ	$QS\,E$	$S\nu/\gamma^2Q$	γQp	$\gamma Q\lambda$	γ^2Qm_I/S	γ^2Q^2h

The speed of light decreases by a factor of $A_P/A_{P'}$ for O after the source has changed its position. This result is obtained by simply replacing A_P by $A_{P'}$ in the general expression for this quantity in Table 4 (second column). The light frequency decreases in the same proportion, whereas the wavelength remains constant throughout. The value of Planck's constant does not change either, and thus the energy decreases in direct proportion to the frequency while the momentum stays the same. The variation of the photon energy is independent of the orientation of the light velocity to the gravitational field.

All the above results can be obtained from those given first by simply replacing A_P by $A_{P'}$. The situation is different if the light is emitted from point P and is later observed by O at point P' in the gravitational field, however. The energy is unchanged during the free fall of the light, whereas the speed of light decreases by a factor of $A_P/A_{P'}$. The value of Planck's constant is determined by the relative speed of the *light source* only, so the frequency also does not change as the light descends (*i.e.*, the value in the fourth column is independent of $A_{P'}$). In order to satisfy the local "$c = \lambda\nu$" condition, it is necessary for the wavelength of light to gradually decrease during its fall by a factor of $A_P/A_{P'}$. This, in turn, means that the momentum of the light increases by the reciprocal of this factor (see fourth column in Table 4). It may seem strange that the momentum of the light increases while its speed decreases, but this is required by the local "E = pc" condition. A similar situation has been noted when light passes through condensed media[95] and will be considered in detail in Chapter 13.

Thus far in the discussion, three cases have been considered: a) the light is emitted from a source fixed at a higher gravitational potential than the observer, b) the light source is moved to a different position in the field prior to emission, and c) the light itself is in free fall from a source fixed at a higher gravitational potential than the observer. A fourth case is considered in Table 5, namely when the source itself is in free fall as it emits light.

It is assumed that the frequency and wavelength of the emitted light have the values ν and λ in the observer's rest frame when measured *in situ* at his location in the gravitational field. The corresponding values for the speed of light, its energy, momentum and inertial mass are c, E, p and m_I, respectively, as in Table 4. The source has been accelerated so that its clock-rate parameter is $\alpha_{M,}$ and the corresponding kinetic scaling factor is, therefore, $Q = \alpha_M/\alpha_O$. It is also located at a higher gravitational potential, with the gravitational scaling factor $S = A_O/A_P$ (see Table 4). The *in situ* values do not change as a result of the light source's movement, in accord with the RP.

The second row of Table 5 gives the corresponding values for these quantities as measured by the stationary observer (O) before the source starts to fall. The energy of the light for O is now QS E, whereas the frequency is $S\nu/Q$ because the value of Planck's constant for light generated from the moving source is Q^2h, as discussed in Section VII. C [eq. (98)]. The speed of light is Sc because of the difference in gravitational potential of the source; the value of the wavelength is $Q\lambda$. The momentum is thus "h/λ" = "E/c" = $Q^2\ h/Q\lambda = QSE/Sc = Qp$. The inertial mass is obtained as "E/c^2" = "p"/"c" = Qm_I/S.

When the light source falls, it is necessary to apply eq. (132) to the scaling factors. Because of energy conservation, as discussed in Section X. A, the kinetic scaling factor Q increases by the same amount (γ) as the gravitational scaling factor (S) decreases. The instantaneous values of the *in situ* light properties are not affected by the free fall of the source, in accord with the

RP. The corresponding values for O are obtained by making the substitutions of S/γ for S and γQ for Q relative to his initial measurements. The observed frequency becomes $S\nu/\gamma^2 Q$, for example, whereas the wavelength becomes $\gamma Q\lambda$.

The quantum mechanical relationships "$E = h\nu$" and "$p =$ "h/λ" are satisfied in all four cases in Table 5. In addition, the mass/energy equivalence relation holds throughout as well ("E" = "$m_I c^2$" = "pc"). As a result, all property values are determined from the speed of light and its corresponding energy and momentum as determined by the observer. This state of affairs is not restricted to light, however. The same relationships hold for all particles of matter, although the Planck energy/frequency relation is of little practical significance for particles such as the electron because of their finite mass and correspondingly high rest energy. The energy of "massive" particles is no longer equal to the product of momentum and the speed of light, but "$E = m_I c^2$" continues to be valid for them. Their velocities between fixed points in a given inertial system also scale in the same way as the speed of light.

Chapter 13

Light Refraction

Light refraction has had a great impact on the development of physical theory over a period of several millennia. It is a phenomenon that is easily observed with the naked eye and yet it took until the early 17th century before it was first possible to formulate a mathematical expression (Snell's Law of Sines) that successfully described it on a quantitative basis. Shortly thereafter, Newton used light refraction to illustrate his Second Law and to support his corpuscular theory of light. However, his views clashed with those of Huygens and other proponents of the wave theory of light, especially in that the two theories led to opposite predictions for the change in the speed of light as it enters water from air.

In the following discussion it will be shown that previous evaluations of Newton's theory have overlooked some very positive aspects. Its failure to correctly predict the decrease in the speed of light in water is not actually proof of the inadequacy of his assumption regarding the composition of light in terms of particles. Rather, it was his inability to make a clear distinction between the momentum and speed of his presumed corpuscles (photons). It can therefore be said that Newton's error was actually a consequence of his thoroughly understandable lack of knowledge of relativity theory. To begin this discussion, a review will be made of the salient features of the competing wave and particle theories and how they led to their opposing predictions of the speed of light in water.

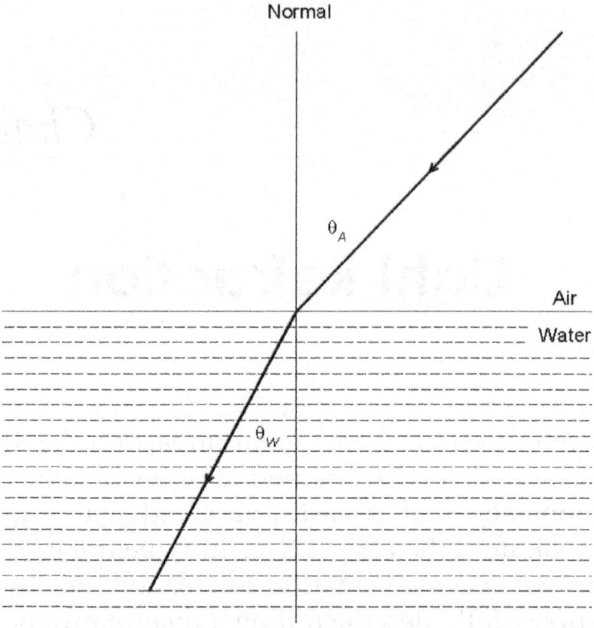

Fig. 7. Diagram showing the refraction of light at an interface between air and water. The relation between the angles of incidence θ_1 and refraction θ_2 in terms of the refractive indices n_i (Snell's Law of Sines) of the two media was viewed by Newton as a clear application of his Second Law of Kinematics, according to which the component of the photon momentum p_i parallel to the interface must be conserved.

A. Comparison of the Corpuscular and Wave Theories

Newton's theory of light refraction[112] was strongly influenced by the work of Snell, particularly the latter's Law of Sines which established the relationship between the angles of incidence and refraction of light rays as they pass between two different transparent media (see Fig. 7):

$$n_1 \sin\Theta_1 = n_2 \sin\Theta_2, \tag{158}$$

where n_1 and n_2 are the refractive indices of the two media.

Newton argued that light consists of particles that are subject to his Second Law. It was assumed that light rays travel in straight lines within each medium and therefore that there are no unbalanced forces in either region. By further assuming that light refraction is caused by a force $\mathbf{F} = \mathbf{dp}/dt$ normal to the interface, he concluded that the momentum p of the particles in a tangential direction must be conserved and therefore that the following equation must be satisfied:

$$p_1 \sin\Theta_1 = p_2 \sin\Theta_2, \tag{159}$$

which is obviously similar in form to eq. (158). Comparison of the two equations thus leads to the following proportionality between the momentum of the particles and the index of refraction of the corresponding light rays:

$$p_1/p_2 = n_1/n_2 = \sin\Theta_2/\sin\Theta_1. \tag{160}$$

Although eq. (160) only deals with momentum (a term not mentioned in *Opticks*[112]), it was used by Newton to make his famous prediction about the speed of light in water. He concluded that since the index of refraction for water is greater than that for air, it must follow that the speed of light must be larger in water as well. This conclusion gained increased significance at the time because it placed his corpuscular theory of light in direct conflict with the wave theory of Huygens and others on this question.

In the wave theory of light it was assumed that Snell's Law of Sines implies that the speed of light decreases as it passes from air into water. The change in angle could be explained[113] by assuming that the distance separating spherical wave fronts decreases as the light passes into a region of higher index of refraction (Fig. 7). According to this model, the wavelength $\lambda = 2\pi/k$ of the light waves changes in direct proportion to $\sin\Theta$, with the result:

$$\lambda_2/\lambda_1 = k_1/k_2 = n_1/n_2 = \sin\Theta_2/\sin\Theta_1. \tag{161}$$

At the same time, it was assumed that the frequency $\nu = \omega/2\pi$ is completely independent of the refractive index for a given medium, so that the corresponding speed of light c_i must be inversely proportional to n_i, *i.e.*:

$$c_i = \lambda_i \nu = \omega/k_i = c/n_i, \tag{162}$$

where c is the speed of light in free space (for which $n = 1$ by definition).

B. Hamilton's Canonical Equations and the Group Velocity of Light

In 1850 Foucault measured the speed of light in water and it was clear that his results stood in irreconcilable contradiction to Newton's prediction. This was quite generally accepted as a complete victory for the proponents of the wave theory of light. The trend had been in this direction anyway since the interference experiments of Young[114] were reported in 1802, which gave strong support to the superposition principle introduced by Huygens in the late seventeenth century, and any remaining question about the wave theory of light was seemingly eliminated in 1864 with the publication of Maxwell's electromagnetic theory. Yet only 40 years later, Einstein's interpretation of the photoelectric effect[115] in terms of light quanta with $E = h\nu$ disturbed this consensus. Then in 1923 Compton was able to construct a quantitative theory for the scattering of x-rays by valence electrons of atoms[116] by using conservation of energy/momentum arguments which were quite consistent with Newton's corpuscular theory.[112] In recognition of these developments, Lewis[117] coined the word "photon" to denote a light quantum, and in succeeding years single-photon detectors have become commonplace in the modern physics laboratory. In spite of this, Newton's erroneous prediction of the speed of light in dispersive media is thought to be irrefutable evidence against a strictly particle theory of light.[118] Instead, the concept of duality[49] is widely used to describe the fact that all matter seems to behave as if it is composed of either waves or particles depending on the type of experiment to which it is subjected.

Yet in later years more accurate experiments[20,85,86] showed that the prediction of the wave theory of light in eq. (162) is not completely accurate either. Instead, the following dependence of the light speed v_g on the derivative of the refractive index with respect to wavelength was indicated:

$$v_g = c/n_g = c/n + (\lambda c/n^2)\, dn/d\lambda. \qquad (163)$$

The presence of the correction term on the right-hand side of this equation has been justified[19] in terms of dispersion effects that are expected to occur when light enters a different refractive medium. Application was made of Rayleigh's theory of sound[18] and its explanation of how beats arise when waves of slightly different wavelength are allowed to interfere.

It is important to return to the question of the assumed dispersion effects, but before doing this, it is instructive to consider how Newton could have so misjudged the effects of light refraction on the speed of light. With centuries of hindsight, however, it is not difficult to find other indications that Newton was on the right track after all. Primary among these is the conclusion that results when eq. (160) of the particle theory is brought into connection with eq. (161) of the wave theory. The latter predicts that the wavelength of light is inversely proportional to the refractive index of the medium, whereas the former concludes that the momentum of the associated particles of light is directly proportional to the same quantity. Combination of these two theoretical relationships leads directly to another, namely:

$$p_1/p_2 = \lambda_2/\lambda_1 = k_1/k_2 , \tag{164}$$

which in turn can be reformulated in terms of a specific proportionality constant:

$$p = h/\lambda = (h/2\pi)k. \tag{165}$$

Stark[119] was the first to arrive at eq. (165). He was influenced by a meeting in which Planck's radiation law was a key topic of discussion.[120] The proportionality constant h in eq. (165) is the same [see eq. (101)] as Planck[121] used nine years earlier to introduce his quantum hypothesis and the corresponding relation between energy E and frequency ν:

$$E = h\nu = (h/2\pi)\omega. \tag{166}$$

Its present-day value is 6.625×10^{-34} Js.

It is a matter of historical fact[119] that eq. (165) was not deduced on the basis of studies of light refraction, but the fact is that this critical result could readily have been achieved over 200 years earlier by simply combining eq. (160) with eq. (161). The reason that Newton and Huygens did not make this connection is most probably because they did not recognize the validity of the other's model for the composition of light and thus were not disposed to making use of any of its respective predictions.[95] The fact that eq. (165) can be derived in a straightforward manner from the two "opposing" theories of light refraction is nonetheless a key observation in theoretical physics. It shows that Stark's recognition of the relation between the momentum of particles of light and the wavelength of the corresponding radiation actually *serves as an important confirmation of both theories*, in particular that of Newton, since it has often been claimed[118] to have been contradicted by the experimental data for light refraction.

None of the above changes the fact that Newton did make a critical error in predicting that the speed of light is greater in water than in air. The reason was *not* his corpuscular theory, however. Rather, it was his inability to compute the light speed in a manner which was consistent with his Second Law. The decisive impulse in this direction was provided by Hamilton and his canonical equations of motion, published 130 years after Newton's *Opticks*. The key equation in the present context is eq. (83). It has been discussed in connection with the derivation of Einstein's famous mass-energy equivalence relation in Chapter 7. B. Perhaps ironically, eq. (83) can be derived in a straightforward manner[122] on the basis of Newton's Second Law and the definition of work/energy as $\mathbf{F}\cdot\mathbf{dr} = \mathbf{dp}\cdot(\mathbf{dr}/\mathrm{dt})$.

The first step in arriving at the correct dependence of the speed of light in refractive media is to apply eq. (83) to obtain the relationship between the energy and momentum of the particles of light in free space. Newton had shown in *Opticks*[112] that white light is decomposed into its component colors when it passes through a glass prism. He concluded that this phenomenon is caused by the varying accelerations experienced by the particles of light associated with different colors when they enter a refractive medium. Since white light travels great distances from the sun without undergoing an analogous decomposition, it follows by the same reasoning that the speed of light has the same constant value c for all corpuscles/photons in free space. As a result, the desired relation can be obtained by integration of eq. (83) for the special case of v = c while setting the constant of integration to zero, namely as:

$$E = pc. \qquad (167)$$

The same equation was used by Stark[119] to derive eq. (165) from Planck's radiation law in eq. (166).

The next step is to generalize eq. (167) for the case of photons in a medium of refractive index n. Because p is proportional to n in Newton's particle theory [eq. (160)], it follows that p can be replaced by p/n without changing the original equality relationship. Furthermore, because of its association with the color of the light, the energy E of the photons can reasonably be assumed to be unaffected as they pass between different refractive media. Consequently, the most straightforward choice for the generalized version of eq. (167) is:[95]

$$E = pc/n. \qquad (168)$$

The speed v of the photons is then obtained from eq. (83) as:

$$v = c/n_g = dE/dp = c/n - (pc/n^2)dn/dp, \quad (169)$$

which is seen to be equivalent to eq. (34). Substitution of the $p = h/\lambda$ relation of eq. (165) then leads directly to the observed expression for the velocity of light given in eq. (163). The corresponding formula for the group refractive index n_g in eq. (169) can be obtained directly as:

$$n_g = d(pc)/dE = d(nE)/dE = n + E(dn/dE). \quad (170)$$

It is also worth noting that Planck's radiation law in eq. (166) is obtained by combining eq. (162) of the wave theory of light with eq. (168):[95]

$$E/p = c/n = \omega/k = \lambda\nu. \quad (171)$$

Substitution of the relation between momentum and wavelength in eq. (165) gives:

$$E/p = E\lambda/h = \lambda\nu, \quad (172)$$

which upon cancellation and rearrangement yields eq. (166). Thus, the revolutionary concept of quantization in Planck's law is also already clearly present in Newton's particle theory of light. Planck based his discovery on a statistical treatment of the entropy of blackbody radiation,[121] but it follows directly from Newton's corpuscular theory when Hamilton's eq. (83) is used to compute the speed of light in dispersive media and thus arguably could have been known as early as 1834, well before the dawn of the quantum age. Had this occurred, the aim of future experiments would have been much more clearly defined than it was historically, namely to accurately measure the value of h in eqs. (165-166).[95]

In summary, when one uses the correct definition for velocity (Hamilton's equation [eq. (83)]), Newton's corpuscular theory is found to be in quantitative agreement with experiment, including most especially with Foucault's determination of the speed of light in water. Newton's error is seen to be his implied assumption that the inertial mass $m = p/v$ of the corpuscles/photons is the same in all media. By contrast, the value obtained from eq. (168) for p and eqs. (163,169) for v is:[95]

$$m = p/v = (nE/c)/(c/n_g) = nn_gEc^{-2} = nn_gh\nu c^{-2}. \quad (173)$$

The inertial mass of Newton's corpuscles increases as the square of the refractive index, which means it is roughly 1.7 times larger in water than in air, causing him to overestimate the speed of light in the former medium by this factor.

C. Light Dispersion in the Wave Theory

After the measurement of the speed of light in water had been made, it became necessary to understand why the value of $c_i = c/n_i$ in eq. (162) estimated using the original wave theory required the wavelength-dependent correction shown in eq. (163). Rayleigh[123] pointed out that the distinction could be explained on the basis of a well-known characteristic of sound whereby waves of slightly different frequency $\omega + \Delta\omega$ and $\omega - \Delta\omega$ are superimposed. The result is a succession of wavelets with variable amplitude.[19] The corresponding wave function is:

$$\Psi = 2A \cos(\omega t - kx)\cos(\Delta\omega t - \Delta kx). \quad (174)$$

The amplitude distribution curve moves with speed $v_g = \Delta\omega/\Delta k$, which is referred to as the group velocity. For light waves in a condensed medium, $\omega = kc/n$, which upon differentiation leads directly to eq. (163).

To justify this approach, it is necessary to assume that whenever monochromatic light from free space enters a transparent medium, *i.e.* where no absorption occurs, a) waves of slightly differing ω and k values are always formed and b) it is the speed of the resulting wave groups that is determined in experiments designed to measure the speed of light in the medium. As discussed in earlier work,[95] this interpretation of the correction term in eq. (163) making use of an analogy to sound waves raises significant questions. For example, it needs to be recognized that, unlike the case for sound, the $\Delta\omega$ and Δk quantities have never been observed for light waves. This conclusion implies that such differences are just too small to be detected, but that raises another question: *how can one measure the speed of the wave groups in actual experiments when their period and wavelength are essentially infinitely long?* Moreover, it also needs to be explained why the frequency and wavelength of the monochromatic light are observed, but their corresponding (phase) velocity indicated in the other factor in eq. (174), *i.e.* $c_i = c/n_i = \omega/k$ in eq. (162), is never measured in refractive media.

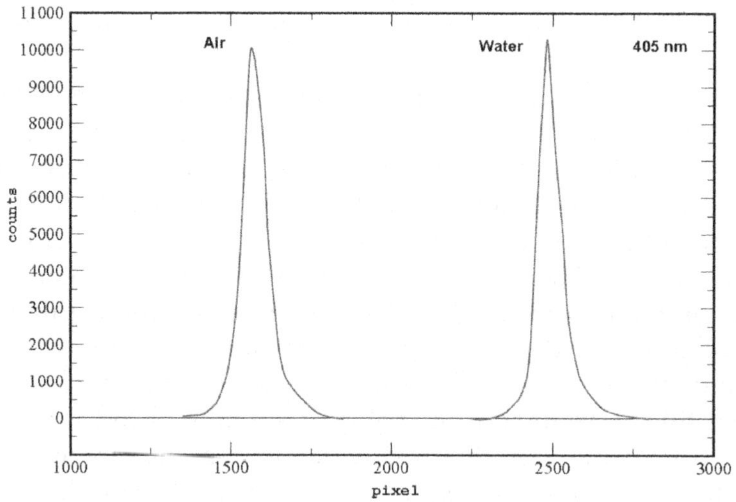

Fig. 8. Distribution of photon counts as a function of pixel time slice for 405 nm light with the test cylinder filled alternately with air (left-hand peak) and H_2O (right-hand peak) and using the long track (see Fig. 2 of Ref. 124). The magnitude of the shift required to bring the peaks into maximum coincidence is used to obtain the difference in elapsed times for light to travel through the cylinder (DL = 0.9455 m) in the two cases and hence the ratio of the velocities of light in air and water.

The analogous situation is unlike any of the classical applications of Rayleigh's theory to sound and water waves. When two musical instruments are slightly out of tune, both the average tone and the characteristic beat frequency are easily audible. When a rock is dropped into a pond, both wavelets and wave groups are clearly visible. In short, the supposed "dispersion" of monochromatic light waves as they enter a refractive medium has never actually been observed and may in fact be purely hypothetical.

In the particle theory, the assumption is that all the photons in monochromatic light of a given frequency have exactly the same energy and momentum in any medium. The second term in eq. (169) comes directly from their momentum dependence within a given medium because of the definition of the light speed in terms of the derivative in eq. (83). There is no requirement that photons of different frequency somehow must be generated because of the interaction with the refractive medium. The model simply assumes (Fig. 7) that the momentum value increases with refractive index n, but that the energy of each photon is not altered as it passes from one transparent medium to another.

Experimental support for Newton's particle model comes from observations of time-correlated single-photon counting (TCSPC). Muiño *et al.*[124] measured the speed of light in water and in air using this technique. Statistical distributions of the photons have been obtained as a function of their time of flight through a given distance. The pattern of these distributions is very similar in the above two media (see Fig. 8); more results of this nature may be found in Ref.[124]. It simply takes longer for the maximum in the distribution to advance through water than it does for air, and on this basis an accurate value of the ratio of the light speeds in the two media was obtained which is in good agreement with the value expected from eq. (163). This experience indicates that individual photons are uniformly slowed as they move into a region of higher index of refraction, and that they all continue to move at the same slower speed as long as they are present in the refractive medium. In summary, the TCSPC experiments[124] show no evidence for the dispersion of light in refractive media that is claimed in the modified wave theory.[18,19] On the other hand, they are quite consistent with the Newtonian view that all photons in monochromatic light are completely indistinguishable regardless of the condensed medium in which they are currently found.

D. Interpretation of de Broglie's Wave-particle Duality Principle

The above discussion shows that the particle theory of light is quite capable of describing light refraction on a quantitative basis, contrary to what has been previously assumed because of Newton's failure to predict the speed of light in water. The question thus arises whether it is possible to explain all experimental observations of light phenomena in terms of such an atomistic theory. The prevailing view among physicists is that there is a wave-particle duality such that matter behaves as particles in some experiments and as waves in others. This approach was introduced by de Broglie[49] and can be viewed as a compromise between the two theories, but duality is not an intuitive concept and it precludes concrete experimental verification.

It is possible to make a different interpretation of de Broglie's principle, however, which is far more consistent with Newton's atomistic views.[95] It relies primarily on the association of a wave with a statistical distribution of many particles: *Some experiments are so precise (photoelectric and Compton effects, the refraction of light and single photon counting) that they reveal the elementary nature of matter in terms of particles, while others (interference and diffraction as the primary examples) are only capable of giving information about the statistical distribution of particles in space and time.*

Support for this statement of the duality principle is found in the Born interpretation[125] of the quantum mechanical wave function Ψ. The absolute square of this function plays a similar role as the intensity in the wave theory of light. In this view, $\Psi\Psi^*$ is a statistical distribution function that gives highly reliable information about large samples of a given entity such as an electron or photon, but is incapable of providing detailed information about the current location of any one of them. For example, in an interference experiment, if the intensity of the beam is small and detection is made with a device such as a photographic plate, the distribution observed early in the counting procedure will vary significantly from one trial to another. However, if the experiment is continued for a sufficiently long period of time in each case, the pattern of detected objects will always stabilize to agree completely with quantum mechanical predictions based on the $\Psi\Psi^*$ magnitude for the associated microscopic system. Most importantly, experiments of this type[126] demonstrate that if the intensity is lowered sufficiently, it is always possible to detect single particles on an individual basis, both for electrons and for photons. This is probably the strongest argument for an atomistic theory of matter and the association of statistical distributions of particles with a quantum mechanical wave packet.

In this view, a single atom, electron or photon is *not* vibrating with a definite frequency and wave length. Rather, the values of λ and ν in eqs. (165-166) are the parameters in Ψ that specify the statistical distribution that many identical particles of this kind possess as an ensemble. As in other applications of statistics, *the corresponding distributions may be quite inadequate for predicting the behavior of individuals, but they provide an unerring guide for trends within very large populations.* The quantum mechanical wave packet thus bears the same relationship to a particle as the histogram does to a member of a sample whose statistical distribution it represents. The latter is a real object, whereas the histogram is only a mathematical abstraction. A light wave is certainly real, but in analogy to an ocean wave containing many water molecules, it is a collective body whose elementary constituents are single photons.

E. Comparison with Gravitational Deflection of Light Waves

The speed of light is also altered when it passes through a gravitational field (see Chapter 9), and thus it is of interest to compare this phenomenon with light refraction.[84] One significant difference is that there is an abrupt change in speed when light passes through an interface separating two different transparent media (Fig. 7), whereas the acceleration is much more gradual when it passes by a massive body. One might conclude from this difference that light rays follow a curved trajectory in the former case, as opposed to the sharp change in slope that occurs when they enter water from air. The evidence from the calculations[78,83] presented in Chapter 9. B speaks strongly against this view, however (see Fig. 5). The angle of displacement was computed in Einstein's work,[76] as well as in Refs.[78,83], not the actual trajectory of the light rays. The key assumption is that the speed of light c′ decreases as the rays pass closer to the sun.[78] Huygens' principle [eq. (125)] has been used to evaluate the angle of displacement by integration over the path between the star at infinity and the earth.

Experimental evidence for the variation of the speed of light with distance from the sun was obtained by Shapiro *et al.*[127] Radar signals were transmitted from earth to either Mercury or Venus and echoes were detected that were retarded by solar gravity. The time delays were expected to increase by almost 2×10^{-4} s when the radar pulses pass near the sun.[95] The predicted increase in echo times was observed to within experimental error.[127] The corresponding decrease in the speed of light c′ in eq. (125) is also expected to be responsible for the observed displacement of star images during solar eclipses, again as discussed in Chapter 9. B.

In summary, there is both experimental and theoretical evidence that light always travels in a perfectly straight line in both refractive media and free space, no matter how strong the gravitational field might be. The change in direction caused by light entering a different refractive medium is due to the fact that the momentum of the photons must be conserved at the interface. The displacement of star images during solar eclipses is explained by the accompanying rotation of the wave front of light as it passes near a massive body such as the sun. An experiment to test the latter interpretation has been outlined in Chapter 9. C. It is assumed that the ratio of distances that the upper and lower rays shown in the diagram of Fig. 6 travel in the period of time in which they are present in different media is exactly the same as that of the corresponding light speeds. If one of the media is free space, then the latter ratio is just the group refractive index n_g of the dense medium.

The difference between the two angles Φ' and Φ'' gives a direct measure of the deviation of n_g from the conventional refractive index n. The exact relation obtained using trigonometric identities is given in eqs. (127) and (128). If $\Phi' = \Phi''$, it follows from both equations that $n_g = n$. The latter equality was the underlying assumption of the original wave theory which led to its erroneous prediction of the speed of light in water as the phase velocity c_i given in eq. (162).

The method described in Chapter 9. C constitutes a direct determination of the group refractive index for monochromatic light.[83,84] It does not require a series of measurements involving different wavelengths of light that allows the determination of $dn/d\lambda$ in eq. (163). It is therefore also a method for determining the speed of light in refractive media. It has distinct practical advantages over the classical techniques[20,85,86] first used to determine this quantity, as well as the more recent procedure[124] involving single-photon counting discussed in Chapter 13. C.

There are also significant reasons of a more fundamental nature for applying this method for determining light speeds. First of all, it is expected to provide valuable new physiological insight concerning the way in which the human eye and other mechanical devices determine the direction from which the images of objects are perceived to come. At the same time, it should provide verification for the conclusion of Chapter 9. B that light rays are not actually bent as they pass by a massive body. Instead, they travel in perfectly straight-line trajectories with varying speeds depending on their separation from the body, thereby causing a rotation of the associated wave front.

Finally, it is important to recall that light refraction also provided the first experimental indication that the speed of light does not depend on the state of motion of the source. Fizeau's experiments with the Fresnel drag effect for light moving in a refractive medium[4] determined the relationship in eq. (31). Substitution of n = 1 leads to eq. (35), as already pointed out in Chapter 3. C. It shows that the speed of light in free space is the same for all values of v, although this aspect of Fizeau's experiment has received little or no attention until the present time.

Chapter 14

Faster-than-c Photons

One of the most famous predictions of STR[5] is that no object can move faster than the speed of light in free space. It is argued that such an occurrence is ruled out theoretically because it would violate Einstein causality, that is, the principle that the time-order of events must be the same for all observers. However, the latter conclusion is based on the LT, which has been shown to lack internal consistency and therefore to be invalid as a physical theory. Opposite predictions for the lengths of objects moving at high speed relative to the observer result from application of the FLC as from taking account of time dilation in conjunction with the light-speed postulate (clock riddle[32]; see Chapter 4. B). A review of the assumptions made in Einstein s original derivation of the LT has shown that it is possible to avoid this contradiction while still satisfying his two postulates of relativity. An amended version of relativity theory has therefore been presented (Chapter 6. A) that is based on the alternative Lorentz transformation (ALT) and remains consistent with the relativistic velocity transformation (VT). The consequences of this change in relativity theory with regard to the possibility of faster-than-c (super-luminal motion) are discussed below.

A. The LT Argument about Einstein Causality

Soon after Einstein's original paper on the special theory of relativity (STR),[5] questions began to arise as to the maximum speed that objects can attain. As discussed in Chapter 3. C, it was pointed out that the group refractive index of light [n_g; see eq. (163)] is less than unity for condensed media in wavelength regions of *anomalous dispersion*,[61] which would mean that the corresponding light speed exceeds c under these circumstances. Since the speed of light in condensed media was found experimentally[20] to be equal to cn_g^{-1} for normally dispersive liquids (with $n_g > 1$), it was argued that light speeds exceeding the value in free space might be possible in media with $n_g < 1$. Sommerfeld[61] vigorously denied the latter possibility on the grounds that it would contradict a basic tenet of STR: Einstein causality. He claimed instead that the speed of energy transport of the waves[19] was the only quantity of experimental significance, and that its value must necessarily be less than c in all conceivable situations.

The relationship between measured time intervals Δt and $\Delta t'$ for the same two events as determined by observers in different inertial systems S and S′ follows directly from the LT [see eq. (17)]:

$$\Delta t' = \gamma(\Delta t - v\Delta x c^{-2}). \tag{175}$$

In this equation, Δx is the distance between the locations of the two events in S and v is the relative speed of S and S′. Consider the case when the two events refer to the passage of an object between two points, so that $u = \Delta x/\Delta t$ is the speed of the object. Rearrangement of eq. (175) this gives:

$$\Delta t' = \gamma\Delta t\,(1 - vuc^{-2}) = \gamma\eta^{-1}\Delta t, \tag{176}$$

i.e. with $\eta = (1 - vuc^{-2})^{-1}$. As long as both v and u are less than c, the signs of Δt and $\Delta t'$ will always be the same and there will be no violation of Einstein causality. The situation changes fundamentally if the object moves with $u > c$, however. In that case it is theoretically possible based on the LT that even if the relative speed of S and S′ is less than c, which it is assumed will always be true, the product of u and v may still exceed c^2 with the result that $\eta < 0$ in eq. (176), thereby constituting a violation of Einstein causality. In other words, the observer in S would find that the object passes point A before it arrives at point B, whereas his counterpart in S′ finds just the opposite time-order of the two events.

Eq. (176) is also the basis for the claim that two events that are simultaneous for one observer will generally not be simultaneous for the other (see Chapter 4.D). It is often claimed that this lack of simultaneity is an inescapable characteristic of the relativity postulates. Clearly, if $\Delta t = 0$ for the two events but both v and

Δx are not equal to zero, it follows from this equation that $\Delta t' \neq 0$. It is important to see that there has never been any experimental verification of "remote non-simultaneity." Indeed, there is a counter-example in the way the Global Positioning System (GPS) technology is designed.[33,105] It is assumed that the time that a light pulse is emitted on a satellite is exactly the same for an observer there as for his counterpart on the earth's surface. One simply has to correct for differences in the rates of clocks located at the two positions in order to determine the actual time of flight on a given clock. The key point is that one assumes that there is a *strict proportionality* between elapsed times measured in the two rest frames, and therefore ignores any effect of the space-time mixing implied by eq. (176).

The experimental indications of faster-than-c motion suggest strongly that one should take a closer look at the theoretical arguments that appear to rule it out. The VT of eqs. (29-30) is obviously relevant to the general question of faster-than-c motion of particles. The discussion in Chapter 3. C has shown that there are three possibilities [see eq. (38)]. First of all, if $u = c$, then $u' = c$ as well, consistent with Einstein's light speed postulate. In the normal case in which $u < c$, it is found that u' is also less than c (with $v < c$). On the other hand, in the case of special interest, if $u > c$ then u' will exceed c as well. There is certainly nothing controversial or even surprising about that, although there is one aspect of the latter result that requires further consideration below.

Once the VT is in hand, one is free to impose an additional condition to arrive at a suitable space-time transformation. It is quite easy to avoid any conflict with Einstein causality by insisting that there is *a strict proportionality between elapsed times for the same event* measured by the two observers. Experience with clocks on circumnavigating airplanes[42] indicates that the proportionality constant (Q) can take on any value depending on how fast each of them runs relative to a standard clock located at one of the earth's poles (the faster the clock moves relative to this position, the slower will be its rate). This suggests that a suitable alternative to eq. (176) is eq. (59), which in turn leads to the ALT of eq. (62), as discussed in Chapter 6. A: Because of the proportionality relation of eq. (59), clearly no violation of Einstein causality is possible as long as $Q > 0$; the elapsed times $\Delta t'$ and Δt will always have the same signs. Moreover, there is no need to assume remote non-simultaneity for the two observers since $\Delta t'$ must vanish whenever Δt does. More importantly, the relevance of eq. (176) is called into serious question by the fact that the LT of which it is a part has been shown in Chapter 4. B (clock riddle[32]) to be self-contradictory. Once this is realized, there is no compelling reason for insisting on either remote non-simultaneity or the prescription against $v > c$ speeds. It is evident from the original discussions of Sommerfeld[61] and Brillouin[19] that their objection to the latter is based entirely on the LT.

B. Tachyons and Photons

The dependence of kinetic energy W on velocity in eq. (77) provides another constraint on the maximum speeds attainable. Since $\gamma = (1-v^2c^{-2})^{-0.5}$ approaches infinity in the limit of $v = c$, Einstein concluded[5,128] that "velocities greater than light have ...no possibility of existence." However, implicit in his statement is the assumption that photons are exempt from this constraint because they possess null rest mass ($\mu = 0$). The possibility of faster-than-c speeds brings a new consideration into this discussion because γ then takes on imaginary values. Feinberg[129] coined the term "tachyons" to describe such hypothetical particles. They would have the interesting characteristic of having imaginary kinetic energies even though the corresponding rest energy μc^2 is real.

The above speculation about tachyons rests on the assumption that eq. (77) enjoys general validity. It is easy to show that this is not the case by considering the very practical case of light in condensed media (see Chapter 13. B). When light enters a medium such as water with $n_g > 1$ (Fig. 7), its speed decreases to $u = n_g^{-1}c$,[20] causing the value of γ in eq. (77) to become finite. Accordingly, this would mean that the energy of the photons should vanish, but *experiment indicates that neither v nor E is altered* as a result of the change of medium. Eqs. (167-172) allow for a continuous change of the speed of photons from values less than c in normally dispersive media through a value of c in free space to higher values in media with $n_g < 1$. One is forced to conclude that eq. (77) is not valid for photons except in the limiting case of $n_g = n = 1$ in free space, even though there is considerable empirical evidence that it does hold universally for particles with non-zero rest mass. The latter conclusion is also reinforced by a calcu-

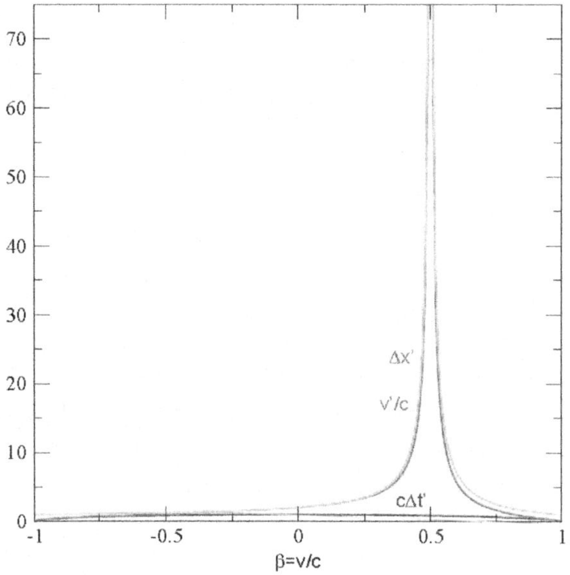

Fig. 9. Diagram showing the variation of the spatial and time intervals $\Delta x'$ and $c\Delta t'$ as a function of the relative speed $\beta = vc^{-1}$ of the observer for a particle travelling at $u = 2c$ relative to the inertial system with $\beta = 0$. Note that $\Delta t' > 0$ throughout, so that no violation of Einstein causality occurs. The variation of the particle speed $u'c^{-1} = \Delta x'(c\Delta t')^{-1}$ is also shown, as calculated from the VT of eq. (177). The values of $\Delta x'$ and $c\Delta t'$ are calculated using the ALT of eq. (178).

lation of the inertial mass of photons in condensed media based on the above equations. From the definition of momentum p as the product of inertial mass m and speed u, it follows that the inertial mass of photons depends in general on both the normal refractive index n and the group refractive index n_g, as shown in eq. (173), which again is different from the classical Einsteinean relation[5] for massive particles ($\mu \neq 0$) of $m = Ec^{-2}$ except for the special case of $n = n_g = 1$ in free space.

C. Adapting the VT and the ALT to v > c Motion

Replacing the LT by the ALT removes the supposed conflict between the Einstein causality principle and faster-than-c motion. It is helpful to consider a specific case in which the observer O′ in S′ views a particle that is moving in S with speed $u > c$ along the x-axis. His measured values for $\Delta x'$, $c\Delta t'$ and $u'c^{-1} = \Delta x'(c\Delta t')^{-1}$ are computed from eqs. (29) and (62) using the proportionality factor $Q = \gamma$ for the specific case of $u_x = 2c$ ($n_g = 0.5$), $\Delta x = 2$ and $c\Delta t = 1$.

The results are plotted in Fig. 9 as a function of the relative speed $\beta = vc^{-1}$ of S and S′. The sign convention is such that $\beta > 0$ when S′ is moving in the same direction as the particle. Qualitatively for each data point, observer O in S sees the particle moving with constant speed $u = 2c$ and observer O′ moving with constant speed v relative to his position.

First of all, it is seen that the $c\Delta t'$ curve has a very simple form because of eq. (59) of the ALT: $c\Delta t' = \gamma^{-1}\, c\, \Delta t = (1 - \beta^2)^{0.5}$. The rate of the clock in S′ is always γ times slower than that in S. This means that the elapsed time measured by O′ has a maximum when $v = 0$, *i.e.*, when S and S′ are not moving relative to one another. More significantly as far as the causality principle is concerned, $\Delta t'$ and Δt always have the same sign. The situation is quite different when eq. (175) of the LT[5] is used (further discussion of this point may be found in earlier work[130]). In that case, $\Delta t'$ vanishes when $uv = c^2$, *i.e.*, when $\beta = 0.5$ in the present example, and it takes on negative values for still greater speeds of S′ relative to S, thereby violating Einstein causality.

Changing from the LT to the ALT has no effect on the way the speed u′ of the particle varies with the relative speed v of the two inertial systems. This is because both transformations are consistent with the VT of eq. (29), as discussed in Chapter 6. A. The fact that $uv = c^2$ for $\beta = 0.5$ in Fig. 9 means that there is a singularity in u′ at that point. In general, the manner in which u′ varies with β according to the VT is counter-intuitive because it means that the faster O′ approaches the particle, the more it appears to speed up from his vantage point. It is easy to understand this result from a purely arithmetical point of view, however. The denominator $(1 - u_x vc^{-2})$ in eq. (29) decreases faster than the numerator factor of c, *i.e.* $(u_x - v)c^{-1}$, as β increases. Analogous behavior occurs whenever $u > c$ in the region $\beta < cu^{-1}$, as can be seen by computing the derivatives of $(u_x c^{-1} - \beta)$ and $(1 - u_x \beta c^{-1})$ with respect to β

Something truly unusual happens when β increases beyond the singular point in eq. (29), however, namely the particle appears to change direction, with u′ abruptly changing from $+\infty$ to $-\infty$. It is easy to avoid this situation, however. One simply has to define η in both eqs. (29) and (62) to be a positive definite quantity, *i.e.* $\eta = |1 - u_x vc^{-2}|$. There is nothing in the derivation of either the VT or the ALT to preclude this possibility on a theoretical basis (see Chapter 6. A), since this choice can be made separately for each value of v. The corresponding equations thus become:

$$u_x' = |1 - vu_x c^{-2}|^{-1}(u_x - v) = \eta\,(u_x - v)$$
$$u_y' = \gamma^{-1}\,|1 - vu_x c^{-2}|^{-1}\,u_y = \eta\,\gamma^{-1}\,u_y \qquad (177)$$
$$u_z' = \gamma^{-1}\,|1 - vu_x c^{-2}|^{-1}\,u_z = \eta\,\gamma^{-1}\,u_z$$

in the case of the VT, and

$$\Delta x' = |1 - vu_x c^{-2}|^{-1} Q^{-1}\,(\Delta x - v\Delta t) = \eta Q^{-1}\,(\Delta x - v\Delta t)$$
$$\Delta y' = |1 - vu_x c^{-2}|^{-1}(\gamma Q)^{-1}\,\Delta y = \eta(\gamma Q)^{-1}\,\Delta y \qquad (178)$$
$$\Delta z' = |1 - vu_x c^{-2}|^{-1}(\gamma Q)^{-1}\,\Delta z = \eta(\gamma Q)^{-1}\,\Delta z$$

in the spatial equations of the ALT. The variation in both u′ and $\Delta x'$ with β in Fig. 9 has been computed on this basis, *i.e.* $u_x' c^{-1} = (2 - \beta)(|1 - 2\beta|)^{-1}$ and $\Delta x' = \gamma^{-1}(2 - \beta)(|1 - 2\beta|)^{-1}$. In this way, the "common-sense" expectation is maintained in the relativistic equations that an object does not appear to change direction until the speed of the observer exceeds that of the object [see also eq. (39) and the discussion in Chapter 3.C]. The latter condition can never occur for faster-than-c speeds of the object, although it is routinely found in every-day experience with corresponding sub-c speeds.

This still leaves open the question of whether the singularities in both u′ and $\Delta x'$ can actually be observed. This would seem to violate the principle that only finite values of physical quantities can occur. Since these are only point singularities, however, one can argue that such an observation is ruled out because of the impracticality of attaining *exactly* the speed at which the singularity should occur. In other words, as β is increased in this region, observer O′ would first find u′ for the object to be increasing to very large values, and then to suddenly start decreasing as the singular value is passed.

These details aside, the main point of the present discussion is that eq. (59) of the ALT leads to an entirely different conclusion about the possibility of faster-than-c motion than does eq. (175) of the LT. The only restriction that remains is that the rest mass of the particles moving with such speeds must be exactly zero. Otherwise, eq. (77) applies and the energy of the particles would increase to infinity as soon as the barrier of $v = c$ is crossed. Since the ALT satisfies both of Einstein's postulates of relativity and avoids the contradiction of the clock riddle[32], there is no longer any reason to doubt that individual photons can travel with $v > c$.

Instead of vigorously pursuing the possibility that the LT is not a valid basis for making such judgments, physicists going back to Sommerfeld[61] have steadfastly denied that faster-than-c motion exists no matter how compelling the positive experimental evidence in favor of the possibility. This starts with the unwillingness to take eq. (163) at face value when $n_g < 1$, even though measurements of the speed of light in media with $n_g > 1$ have consistently been in quantitative agreement with this relation. As discussed in Chapter 13. C, support for this position inevitably falls on unproven assumptions about dispersive forces that supposedly exist in condensed media. Empirical evidence to the contrary is simply unwelcome and easily disregarded by the physics establishment. For example, the data shown in Fig. 8 obtained from the TCSPC methodology[124] give absolutely no indication that the distribution of photons is any different in such media than it is in free space. They indicate instead that each individual photon is moving at the same speed in a given medium and that each one changes its speed in quantitatively the same proportion as it passes an interface between any two such media. This conclusion is perfectly in line with Newton's original theory of light in terms of single photons.[112] It also speaks for a thoroughly straightforward interpretation of numerous experimental studies [62,63,131] that have reported faster-than-c motion of light. It also is consistent with superluminal motion of other particles such as neutrinos as long as they possess non-vanishing rest mass. The fact that Einstein[5] made an unsubstantiated assumption in his original derivation of the LT, as discussed in Chapter 2. C, makes the situation even more transparent. It should not be surprising under these circumstances that one can prove[32] that the LT is self-contradictory based on a comparison of its predictions about length variations using length contraction and time dilation to obtain the results of the theory.

Chapter 15

Summary and Outlook

The discussion of relativity in the preceding chapter is based first and foremost on the proof in Chapter 4.B (clock riddle[32]) that time dilation and FitzGerald-Lorentz length contraction (FLC) are mutually incompatible. Einstein discussed the effect of acceleration on clock rates in his original work.[5] He stated unequivocally that the rates would slow relative to those in the original location, which means that observers in the respective rest frames must always disagree on the value of elapsed times for a given event to occur. Yet the speed of light must be the same for both because of Einstein's second postulate. As a consequence, they must also disagree on the length of a given object because this can be determined by measuring the elapsed time for a light pulse to travel between its two ends (modern definition of the meter).[46] This technique of measuring distances with atomic clocks is well-accepted and is the basis for the GPS navigation technology. An observer O′ traveling with the accelerated clock will measure a smaller elapsed time $\Delta t'$ than his colleague O left behind in the original location, specifically $\Delta t = \gamma \Delta t'$ in the usual notation with $\gamma = (1 - v^2 c^{-2})^{-0.5} > 1$. Multiplication with c then gives $\Delta r = c\Delta t = c\gamma\Delta t' = \gamma\Delta r' > \Delta r'$ for the relationship between the two length measurements. Because of the relativity principle (RP), Einstein's first postulate, it follows that the length of the object *prior to acceleration* must have had the same value ($\Delta r'$) for both O and O′. Since O has remained in his original rest position, this means that the length of the object has changed as a result of the acceleration from a value of $\Delta r'$ to its current value of $\Delta r = \gamma\Delta r' > \Delta r'$. The length of the object has increased by exactly the same factor γ as a consequence of the applied force that caused the acceleration as the rate of the co-moving clock has slowed. Moreover, the increase in length must be exactly the same independent of the orientation of the object because the speed of light is the same in all directions for both observers. In short, according to this argument time dilation in the rest frame of the accelerated observer is accompanied by *isotropic length expansion.* This conclusion is based solely on Einstein's light-speed postulate and the RP.

The results of the Ives-Stilwell experiment[34] are in complete agreement with the above theoretical analysis. They show that the wavelength of light increases with acceleration of the light source at the same time that the frequency declines. Indeed, these measurements obtained the wavelength directly. The conclusion that the light frequency decreases in the process is based solely on the assumption that the product of frequency and wavelength in free space must be equal to the speed of light c independent of the state of motion of the light source, which is clearly in accord with Einstein's second postulate. The RP states that observers traveling with the light source do not notice any change in wavelength, which can only mean that the dimensions of all objects co-moving with them must increase in exactly the same proportion in all directions. The only possible conclusion based on this experience is that isotropic length expansion accompanies time dilation, exactly as deduced from the arguments of the clock riddle.

Opposite results are predicted based on the FLC, depending on the orientation of the objects, so there is a clear contradiction between these two ways of applying STR: $\Delta x = \gamma^{-1}\Delta x'$, $\Delta y = \Delta y'$ and $\Delta z' = \Delta z$. Both time dilation and the FLC result directly from the Lorentz transformation (LT), which therefore proves that something is wrong with its derivation reproduced in Chapter 2. B. Given this situation for such a basic tenet of STR, it is important to be clear about the principles of logic in general. The only rational means of avoiding the above contradiction would be to show that something is wrong with the premises of the argument. This is impossible in the present case because only time dilation and the FLC are assumed in addition to Einstein's two postulates of STR,[5] and the relevant characteristics of these predicted effects have been faithfully applied in the above example. Just because huge numbers of well-informed physicists have concluded over a period of more than a century that there is nothing wrong with Einstein's theory is of no consequence whatsoever in this discussion. There is nothing personal about logical discourse. In short, one must either find a flaw in the argument or accept its conclusions.

Secondly, it should be clear that just because a logical argument leads to many successful predictions for the accompanying theory does not constitute proof of the assumptions used to construct it. Once it is clear that one of the premises is indeed false, it becomes important to find another explanation for the confirmed results of the theory. In the case of STR many of the experimental verifications such as aberration of starlight at the zenith and the Fresnel light-drag experiment are fully explained in terms of the relativistic velocity transformation (VT) of eq. (29) and therefore do not require the LT in any way. The Michelson-Morley[3] and Kennedy-Thorndike[51] experiments are also fully explained by the VT, even though they are often incorrectly claimed to be verifications of the FLC and time dilation.

The above information about the central role played by the VT in relativity theory brings one quickly to the crux of the problem with the LT. Einstein[5] made an undeclared assumption in deriving the LT from the general version of the Lorentz transformation (GLT) of eq. (15). This assumption forced the free parameter ε in those equations to take on the value of unity required to obtain the LT. He gave no explanation for asserting that this function could only depend on the relative speed of the two inertial systems, yet the FLC and Einstein's symmetry principle (ESP) depend totally on this assumption. It is therefore not at all surprising that numerous predictions of the resulting theory are incorrect. On the other hand, any other choice for ε leads to the VT without requiring the LT to be involved in any way. In summary, there are two classes of experimental predictions of STR, those that can be derived on the basis of the VT alone and those that cannot. The clock riddle[32] proves that the LT is invalid and therefore that all the predictions such as the FLC and the ESP that do not rest solidly on the VT are to be treated with the utmost suspicion.

The degree of freedom[132] in the GLT represented in the form of the parameter ε is the key to developing a revised version of relativity theory that removes the contradiction laid bare by the clock riddle. The Fresnel light-drag experiment gives positive support to the light-speed postulate [see eqs. (31) and (35)], as do all the other experiments mentioned above that can be derived exclusively from the VT. In the search for that alternative value for ε, it is quite helpful to consider experiments that were carried out with the transverse Doppler effect employing the Mössbauer effect[37,39,40] in conjunction with a high-speed rotor. Contrary to statements of the various authors, it is clear that the results of these investigations contradict the expectations from STR. In particular, the empirical formula of eq. (49) is not consistent with the ESP since the latter predicts that only a frequency red shift should be observed. If one looks upon both the x-ray source and the absorber as clocks, the results indicate instead that their respective rates slow down in direct proportion to $\gamma(v)$, where v is the speed of the clock relative to the axis of the rotor on which it is mounted. This observation is consistent with Einstein's example[5] of an electron moving in a circular path before returning to its original location. Subsequent experiments with atomic clocks onboard circumnavigating airplanes[42] produced a similar result, whereby in the case the earth's center of mass (ECM) takes the place of the rotor axis as the reference frame from which to compute the speeds to be inserted in the γ factors. Altogether, these results satisfy the Universal Time Dilation Law (UTDL) given in eq. (57). The term "objective rest system" (ORS) has been defined in Chapter 5. C to specify the reference frame in each case, *i.e.* it is the rest frame in which the electron is first accelerated in Einstein's example, it is the axis of the rotor in the Hay *et al.* experiment and it is the ECM for the airplane tests with atomic clocks.

The above experimental data demonstrate that Einstein could have taken a different approach to determine the correct value of ε in the GLT, namely to base this choice on his example[5] of a clock co-moving with an accelerated electron. The UTDL can be brought into a suitable form to do this, namely eq. (59). This conclusion is reinforced by the GPS methodology in which exactly such a proportionality relation is used to adjust clocks on satellites so that they run at the same rate as their counterparts on the earth's surface. The parameter Q in eq. (59) is obtained by taking the ratio of the two γ factors in the UTDL of eq. (57). A typical value is $Q = \gamma$, where one of the rest frames is the ORS. The main characteristic in this way of treating time dilation is that it is *asymmetric*. If the roles are reversed so that the observer is now accelerated, one simply has to replace Q by $Q' = Q^{-1}$ in eq. (59). This is the antithesis of the ESP-relation that results from assuming the LT. If the ORS is known, there is never any doubt as to which clock is running slower than the other. It is the one that moves at the higher speed relative to the ORS.

It is straightforward to determine the value of ε in the GLT once the above assumption has been made. One simply has to combine the proportionality relation of eq. (59) with the time relation of the GLT. The alternative version of the Lorentz transformation (ALT) is then obtained by substituting this value into the GLT, whereby eq. (59) therefore becomes the relation between elapsed times measured in the respective rest frames. The result is the space-time transformation of eq. (62). An equivalent way of obtaining the same result is to meld eq. (59) with the VT directly. This procedure emphasizes the direct relationship between the ALT and the VT and makes it obvious that any experimental prediction such as for the aberration of starlight at the zenith that results from application of the VT is also perfectly consistent with the ALT. The latter clearly satisfies both of Einstein's postulates and bears the same relationship to Maxwell's equations as does the LT. Because of its use of the proportionality relation of eq. (59), the ALT can be more descriptively named the GPS-compatible LT or GPSLT for short. The original LT subscribes to the ESP and therefore cannot be applied for any situation where an asymmetric relationship between clock rates is assumed as in the GPS methodology.

Use of the ALT/GPSLT in relativity theory has wide-sweeping consequences. First of all, it replaces the LT and demonstrates that Einstein's two postulates can be satisfied while avoiding the contradiction in length measurements resulting from the clock riddle of Chapter 4. B. Eliminating the LT from consideration also means that the condition of Lorentz invariance is null and void. It is replaced by eq. (64) and its inverse in eq. (65). A key result in this respect is the identity of eq. (67) that relates the function η in the VT to η' in its inverse (see Appendix B). The obvious symmetry of the Lorentz invariance condition is shown thereby to be non-essential for satisfaction of the RP. Observers in two different rest frames must use equivalent sets of equations but still take account of the fact that if the first one perceives the second moving in the + x direction, his counterpart must find that the "other" is moving in the –x direction.

One of the most critical results of the LT is space-time mixing. Belief in this characteristic has caused generations of physicists to deny the principle of remote simultaneity of events. By contrast, because of the proportionality relationship in eq. (59), the ALT demands that this principle be upheld in all circumstances. The ALT also leaves open the possibility that photons can travel with faster-than-c speeds in condensed media, in agreement with many experimental observations and the definition of the group refractive index, whereas the LT claims that such an occurrence would violate Einstein causality. The FLC is also eliminated from consideration, thereby removing the contradiction of the clock riddle. Claims to the contrary are seen to result from not recognizing that the VT is sufficient in itself to explain the results of the Michelson-Morley and Kennedy-Thorndike experiments and failing to see the difference between relativistic length contraction and the change in the de Broglie wavelength with the momentum of individual particles in a large sample. As noted above, the ALT replaces the FLC of the LT with isotropic length expansion, consistent with time dilation and the light-speed postulate (see Appendix A).

Relativity theory based on the LT espouses a subjective view of the measurement process through its reliance on the ESP. There has never been any experimental verification of the latter. Rather, in experiment after experiment it has been demonstrated that it is always possible to know which clock is running slower, consistent with the UTDL. Sherwin's assertion[41] that the predicted symmetric relationship between clock rates would come into play if only strictly uniform translation is involved has never been confirmed. If one accepts the possibility that an accelerated clock runs slower than one left behind in its original rest frame, it requires a leap of faith to claim that this thoroughly asymmetric relationship will be erased instantaneously just because the clock suddenly assumes a constant velocity. One expects instead that the rate of the accelerated clock simply remains constant after a state of pure translation is attained, no matter how much it was accelerated up to that point. The ALT, by contrast, allows for a completely objective view of the measurement process. The rate of a clock only depends on its speed relative to the ORS. The parameter Q in the ALT is simply a conversion factor between the respective units of time in the two rest frames. Consistent with the UTDL, it remains constant as long as neither of their relative speeds to the ORS changes. This is true even if one or both clocks is subject to circular acceleration, as for example in the case of the transverse Doppler experiment where each of the clocks is mounted on a rotor.

The experience with the measurement of light frequencies and other clock rates suggests strongly that all other physical properties also conform to a strictly objective measurement process. A simple way to express such relationships is through the use of a rational set of units in each inertial system. Formally, one can add the Principle of Rational Measurement to the list of postulates of relativity (PRM). It states that two observers will always agree on the *ratio* of their respective measured values for two quantities of the same kind, even though their individual numerical values differ because of their use of different systems of units. An important goal in relativity theory is accordingly to establish the scaling characteristics of each physical property with changes in the state of motion of either the observer or the object of measurement. It is assumed that the effect of applying a force to an object is to not only produce acceleration in accordance with Newton's Second Law, but also to change the values of its physical properties. Because of the RP a co-moving observer will not be able to notice such changes unless he carries out measurements for objects which are in relative motion to him.

Analysis of confirmed experimental relationships helps to establish the way each property changes. For example, the constancy of the speed of light for all observers is a clear indication that the unit of velocity is the same in all rest frames. This is turn leads to the conclusion that distance measurements scale in exactly the same manner as elapsed times, *i.e.* isotropic length expansion accompanies time dilation. Thus, if the conversion factor for time is Q in the ALT, it is also Q for the corresponding factor for distance but is Q^0 for velocities. Measurements of the way inertial mass of an electron changes with its speed in the laboratory[66] are also instructive. They show that this quantity varies in the same way as time and distance, *i.e.* it also scales as Q. Objectivity then allows one to deduce the conversion factors of other properties based solely of their composition in terms of the fundamental units of distance, inertial mass and time in the mks system. For example, frequencies vary as Q^{-1}, energies as Q (because of the $E = mc^2$relation, for example), and angular momentum as Q^2. The latter scale factor helps us to understand why the energy of light waves increases with acceleration of the light source while their frequency decreases by the same factor. This means the energy/frequency ratio varies as Q^2, consistent with the fact that Planck's constant h has units of angular momentum (Js).

A summary of the conversion factors is given in Table 1. They are all integral powers of the conversion factor Q for elapsed times, which is certainly a simplifying feature for the overall theory. Momentum **p** is a key quantity in this discussion. It is the product of inertial mass m and velocity **v**, and that naturally raises the question of what reference to use to determine the value of **v**. To answer this question it is important to recall the precise meaning of Newton's Second Law. The momentum results from acceleration of the object due to an applied force **F** at a particular point in space. The instantaneous value of **v** relative to this origin is therefore uniquely specified and is the same for all observers regardless of their state of motion (because **v** scales as Q^0). With this specification it is possible to state unequivocally that **p** scales as Q, the same conversion factor as for inertial mass and energy. Confusion results because in non-relativistic physics it is commonplace to assume that the velocity that determines the momentum of an object can be defined relative to an arbitrarily chosen observer regardless of his state of motion. The situation is closely related to what has been discussed above for time dilation under the assumption of the ESP. It has been demonstrated experimentally that it is false to claim that the rate of a clock depends only on its speed relative to the observer. The origin of the applied force that produces momentum serves as the ORS for determining the rates of the accelerated clocks. Therefore, it is only consistent to assume that the velocity **v** used in the definition of the associated momentum **p** is uniquely defined relative to this rest frame. It is important to note that the principle of conservation of momentum therefore applies equally well in all inertial systems. The respective observers will disagree on the values of the inertial masses of the participating objects in a collision and therefore on their momentum values, but the conservation law will be upheld for all according to this scaling procedure.

The above conclusion about which velocity is to be used to determine the momentum of an accelerated object is also relevant to the relativistic theory of electromagnetic interactions. Heretofore, it has invariably been assumed that the velocity **v** in the Lorentz force equations [see eq. (113) in Chapter 8. A] is defined relative to the observer. Einstein[5] assumed for example that $v = 0$ in the rest frame of the accelerated electron. He then showed that Maxwell's equations are invariant to the LT provided the various components of the electric and magnetic fields satisfy the relations in eq. (110) [with $\psi(v) = 1$]. The conclusion thereby is that two observers will measure different values for these quantities depending on their relative speed to one another. The implication is that the above procedure is the only way to satisfy the RP, but that is patently false. As discussed in Chapter 8. D, the RP can be fulfilled in a much more straightforward manner by assuming that the origin of the applied Lorentz force is the reference frame (ORS) for determining **v**, similarly as with the definition of momentum in the previous paragraph. One of the consequences of Einstein's derivation is that the transverse force components measured by the two observers will differ by a factor of $\gamma(v)$, as shown in eq. (114). This conflicts with the result for the scaling of the force components given in eq. (105) which states that both observers must agree on the value of each of the force components. All observers also agree on the values of other properties on an *absolute* basis according to this interpretation, but they will generally disagree on the corresponding numerical values because of the fact they use different system of units to express their results.

The fact is that a perfectly valid relativistic theory of electromagnetism is obtained, as Planck suggested[65] to Einstein shortly after his original work, by combining Newton's Second Law and equating $\mathbf{dp}/dt = d(\gamma\mu\mathbf{v})/dt$ to the Lorentz force $\mathbf{F} = e(\mathbf{E} + \mathbf{v}c^{-1}x\mathbf{B})$, whereby the velocity **v** is defined relative to the origin of the applied force (the laboratory/ORS) in both cases [see eqs. (112-113)]. This equation has been verified to quite high precision in circular proton accelerators[133] under extremely sensitive conditions. The key point is that the LT is not involved in any way in the above derivation, and the observed results are perfectly consistent with the concept of uniform scaling in the revised ALT version of relativity theory.

The RP comes into play when an identical electromagnetic interaction (same **E** and **B** fields) occurs at a stationary origin in a different inertial system S′. Accordingly, whatever data the stationary observer in the original inertial system S measures for the interaction in his rest frame must be faithfully duplicated by his counterpart in S′ for the equivalent interaction in his rest frame. This condition is perfectly analogous to what Galileo described in his original description of the RP. He imagined carrying out simple experiments in the hold of a ship while it was in different states of motion. The passengers were unable to tell the difference between remaining at the dock and sailing on a perfectly calm sea. One can expand on the comparison by envisaging two other observers who are moving relative to the respective origins (ORSs) of the two interactions. If the speed of the first relative to the ORS in S is the same as that of the second relative to the ORS in S′, they will also agree on all corresponding measurements for the two interactions. This is because both sets have the same conversion factor Q between their units and those in the respective ORS.

The above arguments are obviously not specific to electromagnetic interactions. They are applicable to any conceivable type of force. They adhere strictly to the Newtonian principle that there is no such thing as absolute motion, but they go beyond Einstein's STR implementation. It needs to be recognized that observers in different states of motion can nonetheless disagree as to the values of measured quantities because of their unwitting use of different standard units. This state of affairs makes it advisable to augment the original statement of Galileo's RP, as has been done at the conclusion of Chapter 8. D. This includes a specific denial of the STR assertion that the laws of physics must be invariant to the LT in order to satisfy the RP.

There is still one loose end in the above description of the relativistic theory of electromagnetic interactions. The discussion thus far has circumvented the fact that many key quantities such as electric charge are used in this theory that are not expressed in terms of the standard units of distance, time and inertial mass. This fact obviously hinders the full implementation of the uniform scaling procedure for such interactions. It is shown in Appendix D that there is actually a degree of freedom in the definition of electric charge in Coulomb's Law that enables one to overcome this problem in a variety of ways.

Finally, a derivation of the relativistic Doppler effect is given in Chapter 6. D. It demonstrates that the $(1 - vc^{-1}\cos\Theta)$ term is actually not a relativistic effect, but rather simply reflects the fact that the measured distance between successive wave crests depends on the motion of the light source toward or away from the observer. It is therefore not subject to the usual rule for obtaining inverse relationships in relativity theory, for example. The transverse Doppler effect is relativistic in nature and is simply the consequence of the uniform scaling of time and distance between different inertial systems. The corresponding shift in frequency does not satisfy the ESP but rather is a direct consequence of the UTDL of eq. (57).

After Einstein published his STR paper,[5] he began to look for ways to extend the methods he had developed in this area to the treatment of gravitational effects. He introduced his Equivalence Principle (EP) two years later[38] and came to a number of important conclusions. The basic idea underlying the EP is that translational acceleration is indistinguishable from a uniform gravitational field. For example, it is claimed that one can eliminate the effects of **g** by replacing it with an equivalent acceleration in the laboratory of $\mathbf{a} = -\mathbf{g}$.[134]

There are two key results in the 1907 paper that bear close consideration. Einstein used the EP by studying Maxwell's equations to deduce that the energy, frequencies and the speed of light increase as the source is moved to a higher gravitational potential. Since the speed of light in free space is equal to the product of wavelength and frequency, it follows that there is no corresponding change in the wavelength as the gravitational potential is varied. The fractional change is proportional to $S = (1 + ghc^{-2})$, which is reminiscent of the factor Q in kinetic scaling. He concluded on this basis that there should be a bending effect for light passing close to a massive body such as the sun. He later used Huygens' principle on this basis to compute the angle of deflection of light rays coming from infinity to have a value only half as large as the presently accepted result.

The variation of the speed of light with gravitational potential also tells us something about the EP itself, however. When an external force is applied to a light source, the speed of light continues to have a value of c for an observer at the same gravitational potential no matter how great the change in velocity. Yet, when the same light source changes its position in a gravitational field, the speed of the emitted light does change in a predictable manner. That means that it *is* possible to distinguish between translational acceleration and that induced by gravity. *As a result, the EP is not absolute and becomes more of a helpful guide than a true principle.* In terms of uniform scaling, this observation indicates instead that the effects of applying an external force to an object need to be considered separately from those resulting from a gravitational field. There are two easily distinguishable conversion factors, one (Q) depending on the speeds of the two rest frames relative to their common ORS and the other (S) on their respective positions in a gravitational field. This was the position taken by Hafele and Keating[42] in the analysis of clock rates on circumnavigating aircraft.

Only the kinetic factor Q in Table 1 is required to understand the results of the transverse Doppler experiments with rotors.[37,39,40] It is the ratio of the $\gamma(v)$ factors in the UTDL of eq. (57). If the absorber clock is moving faster, then its clock rate is lower and a blue shift is observed, whereas the opposite is the case when the x-ray source is located at the outside of the rotor's rim. On the other hand, only the gravitational factor S has an influence of the frequency measurements in the Pound-Snider experiment[91] since there is no relative motion of the source and detector in this case. The unit of frequency is smaller at the lower potential and therefore an increase in light frequency is observed there compared to the standard value measured by the observer at the position of the source at higher altitude. The absolute frequency is unchanged. In short, the gravitational red shift is completely accounted for by the conversion factors in Table 2 for elapsed times and frequencies. Moreover, it is not correct to say that the photons have weight, since they have null gravitational mass. The energy of the photons is conserved during the free fall, but its value is measured to be greater at the lower altitude because of the smaller value of the unit of energy there.

Schiff's computational method[78] throws further light on the gravitational scaling procedure. His calculated value for the deflection angle for light rays was again half the correct magnitude when only measured times were scaled in accord with the gravitational red shift. He therefore proposed a scheme for distance scaling (referred to as the GFLC in Chapter 9.A) that is very similar to that of the FLC in STR.[5] Distances radial to the gravitational field of the sun are scaled by a factor of S, whereas those transverse to the field are not scaled, consistent with Einstein's above result from the EP. This change in the procedure leads to a doubling of the deflection angle, in agreement with the best result for this quantity obtained from GTR.[76] Analysis shows, however, that Schiff's radial scaling (SRS) only leads to a change in the distance travelled by the light. It has no effect on the path taken by it, only on the speed with which it travels from the vantage point of the observer at infinity; it continues to move at velocity $\mathbf{v} = c$ in a perfectly straight line. If the radial scaling is taken at face value, it must lead to a veering of the light away from its initial direction (see Fig. 5), but taking this approach does not lead to the correct prediction of the deflection angle. The conclusion is that the SRS allows the distance travelled by the light to be computed accurately and that this value is then assumed to be exactly the same as measured by any other observer who is stationary in the rest frame of the active mass (sun). This is consistent with Einstein's 1907 analysis,[38] which shows an equal scaling with gravitational potential by a factor of S for the speed of light and its frequency, and therefore no change in its wavelength. Generalization leads to the gravitational isotropy of space (GIS), as shown in Table 2 by the conversion factor S^0 for distance. The SRS effectively changes the conversion factor for time in Schiff's procedure, causing the local elapsed time to be less than is expected when only the conventional scaling of time is used. Qualitatively, the SRS is equivalent to the "curved space" of the GTR. It is necessary to obtain a quantitative prediction of the trajectory of the light rays, but it is still perfectly consistent with the GIS and therefore Newton's "flat space." It is important in this analysis to clearly distinguish when a computational procedure is a direct reflection of reality from when it is merely a mathematical artifice that guarantees a particular result. The SRS is such an artifice, but the GIS is a true law of physics that is to be taken completely literally.

For this reason, the conversion factors that are directly connected with the SRS are shown with an asterisk in Table 2. They only apply for an observer (O*) in a gravity-free region of space and are used purely to obtain accurate parameters for the trajectories of objects moving in a gravitational field. *The above analysis shows that light rays are not bent by a gravitational field no matter how strong it might be.* The effect of gravity is purely on the speed with which the light travels. As shown in Fig. 4, this leads to a rotation of the light's wave front by an angle which has exactly the same value as the "deflection angle" in the GTR analysis. This result suggests that the way we judge the direction from which light approaches the eye is by drawing a line normal to that of the wave front. An experiment has been suggested with light refraction (Fig. 6) that is capable of testing this physiological assumption.

It has long been pointed out by many distinguished authors that there are only three crucial tests of GTR: the gravitational red shift, the "deflection" of light rays passing by massive objects and the precession of the perihelion of Mercury's planetary orbit. The gravitational red shift is accounted for in Table 2 and is just one of many effects that are governed by uniform scaling. The phenomenon of light deflection has been accounted for with Schiff's methodology,[78] which is an application of uniform scaling as well. The remaining question is therefore whether the precession of Mercury's orbit can be treated successfully using similar computational procedures. In this connection it is noteworthy that Schiff was able to obtain his results without assuming any influence of g-forces on the light's trajectory. This is not quite true, however, since the conversion factor $S(A_p)$ contains the local g value explicitly [see eq. (130)], but at no point in his procedure is the velocity adjusted by adding the usual $g\Delta t$ factor corresponding to gravitational acceleration. The reason lies in the application of kinetic scaling to this problem. It is reasonable to assume that Newton's ISL can always be used to compute the *local* value of g, in agreement with all experimental observations. The ISL is proportional to the gravitational mass M_s of the sun and inversely proportional to the square of the distance r separating the passive mass from the sun/active mass. To obtain the g value to be used by a stationary observer O* in the rest frame of the sun, it is therefore necessary to scale the local value of g to his units. In Table 1 it is assumed that M_s must scale as Q^0 while r scales as Q (the universal gravitation constant G is also assumed to be the same for all observers). As a result, g must scale as Q^{-2} for O*. *But* $Q = \infty$ *for light,* *so that the appropriate conversion factor is zero in this case, and therefore g = 0 for the observer at infinity.*

The fact that there is no g-acceleration in Schiff's calculation of the trajectory of light rays is thus merely consistent with the general kinetic scaling procedure. The situation will be quite different for massive objects such as the planet Mercury, however. In that case, g is not eliminated by kinetic scaling and it is also affected by gravitational scaling. The SRS comes into play in determining the latter conversion factor for O*. Because r* scales as S, it follows that the conversion factor for g must be proportional to S^{-2} on this basis. Since the inertial mass of the sun is equal to its gravitational mass for O* because of his stationary position in that rest frame, it is unclear how to scale M_s in the ISL. Test calculations settle this question quantitatively. Mass is to be scaled as inertial mass, so that there is an additional fact of S^{-1} to be combined with the S^{-2} factor for radial distance scaling. This gives the S^{-3} scaling for g listed in Table 2. The overall conversion factor from local units to those of O* to obtain **g*** is thus $Q^{-2}S^{-3}$.

The calculation of planetary trajectories is carried out by O* in his units for an elapsed time Δt* in each cycle. He combines the scaled initial velocity $\mathbf{v}_i^*$ with **g***Δt* using the VT to obtain the final velocity $\mathbf{v}_f^*$ of the planet at the end of the time cycle. The vector distance **Δr*** traveled by the planet in this period is obtained in the usual way as the average of the initial and final values of the velocity ($\mathbf{v}^{a*}$) multiplied with Δt*. The GIS is then invoked to conclude that the magnitude of the distance traveled in local units is the same as in O*'s units, *i.e.* Δr*. The direction is that of the local velocity $\mathbf{v}_i$, which is obtained by the back-scaling of $\mathbf{v}_f^*$. The resulting trajectory for the planet[79] is in quantitative agreement with that obtained with GTR.

The above procedure can be generalized in a straightforward manner for the interaction of an arbitrary number of massive bodies (Chapter 11. F). Only the gravitational masses, locations and local velocities are required as input. One of the advantages of the method is that its results are completely independent of the choice of origin for the coordinate system. The GIS insures that the locations of the planets and stars are the same for all observers who are stationary in a given rest frame such as the sun's in Schiff's original methodology. The gravitational masses are the same for all observers, even those in relative motion to one another. This means that Newton's ISL can be applied for any pair of objects in the system of units in an arbitrary rest frame. To convert trajectories from one rest frame to another, it is sufficient to know the value of the kinetic scaling parameter Q that connects them. The larger the value of Q, the more the universe appears to shrink for the other observer.

Thus far in the discussion, emphasis has been on the three critical experiments mentioned above. Universally quantitative agreement is found between the predictions of GTR and the uniform scaling methodology in each of these cases. There are other areas in which there is clear disagreement, however. One of the most glaring is the discrepancy between the predictions[29] of the precession rate of orbiting satellites [compare eqs. (154) and (155)]. It is clear from this example that at least one of the methodologies is not completely error-free. It is true that the underlying assumptions of the uniform scaling method are *ad hoc* in nature, but then the situation is no different for Newton's Laws of Motion. The important point is whether the predictions based on such laws are correct or not. One should also avoid the delusion that there is nothing *ad hoc* about GTR simply because of the elegance of its construction. The uniform scaling method has the advantage of simplicity in applying it, and that characteristic is of inestimable value in practical computations. In the last analysis, the only fair way to compare the theories is to carry out definitive experimental tests that clearly differentiate between their respective predictions.

The treatment of gravitational light deflection leads naturally to a discussion of the effects of light refraction in condensed media in Chapter 13. A review of Newton's corpuscular theory[112] shows that it has been widely misjudged because of his failure to anticipate that the speed of light is lower in water than in free space. On the basis of his Second Law, Newton proved that the momentum p of photons is directly proportional to the index of refraction n of the medium. Since the wavelength of light is inversely proportional to n, this result is confirmed by quantum mechanics and Stark's conclusion[119] that $p = h/\lambda$ in free space. Planck's radiation law[121] in eq. (166) could also have been deduced on this basis well before the advent of quantum mechanics by using Hamilton's equation and Newton's experimental result that the speed of light is independent of color/energy.

Newton failed to distinguish between momentum and speed in his work and thus came to the erroneous conclusion that the speed of light should increase when it enters a medium with a larger value of n. Since the energy E of the photons is equal to pc in free space, the above proportionality leads to the general equation of state, $E = pc/n$, for light in condensed media [eq. (168)]. Hamilton's canonical equations allow the dependence of the light speed on n to be deduced as $v = dE/dp$, as shown in eq. (169). The result agrees quantitatively with the empirical formula for light speed c/n_g given in eq. (163). It shows that the corpuscular theory works perfectly if there is no error in its application. Further analysis shows that the inertial mass m of the photons is proportional to the product of n and n_g [see eq. (173)], in contrast to Newton's assumption that m does not change as the light passes through an interface between two media.

A key assumption in Newton's theory is that photons of the same energy all travel with the same speed in a given medium. This contention is supported by the results of TCSPC experiments[124] which show no change in the appearance of the photon distribution in going from one medium to another. This result contradicts the position of the wave theory of light that the distribution is changed by dispersive forces in the different media. One has instead a simple model in which each photon in monochromatic light has exactly the same energy and momentum. The difference between n and n_g is equal to $E dn/dE$ in eq. (170) and arises in a straightforward manner by taking account of the above equation of state for photons in condensed media.

Light refraction has also played a key role in the controversy over whether objects can travel with $v > c$, as discussed in Chapter 14. Since the group refractive index n_g has values less than unity in wavelength regions of anomalous dispersion, it follows from eq. (169) that photons can travel with such "superluminal" velocities. Beginning with Sommerfeld,[61] this possibility has been vigorously denied, even though there have been numerous experimental indications that appear to confirm $v > c$ motion. The arguments against taking the latter results at face value are always based on the supposed effects of dispersive forces. Yet, single-photon counting experiments contradict this position, whereas it should be clearly understood that there is absolutely no direct evidence to support the existence of the dispersive forces.

The real reason that the physics community refuses to accept that photons can travel with faster-than-c speeds is the unconditional belief in the LT, which brings us full circle back to the start of this summary. The LT is self-contradictory because it gives opposite results for length measurements on a moving object depending on whether one uses atomic clocks to make the determinations in conjunction with the light-speed postulate or simply relies on the prediction of the FLC for this purpose. This circumstance rules out any possibility for the LT to be a valid space-time transformation, and thus clearly calls into question any and all conclusions that are based on it unless they can be derived completely independently on some other well-founded basis such as the VT. That is not the case for both the FLC and the denial of superluminal motion, among other well-known predictions of the LT.

Finally, it should be noted that the condition of Lorentz invariance has had applications that go well beyond the realm of classical physics. Physicists have long refused to believe that a potential exists that is capable of binding an electron inside the neutron, even though a decay process for this particle in which an electron does not appear as a product has never been observed. The standard argument is based on the fact that it is mathematically impossible to devise a sufficiently short-range potential to bind the electron in a Hamiltonian that is invariant to a Lorentz E,p transformation. The quark hypothesis is based directly on this assumption since it is assumed that such a potential can exist if the constituents of a nucleus are of much greater mass than the electron. Yet, the fact remains that an electron always appears in the decay of a neutron. More details about this topic are given in earlier work,[135] but the knowledge that Lorentz space-time invariance is founded on an invalid transformation raises questions about the wisdom of basing particle physics on a related condition.

Appendix A

The Giant Twin's Travels

1. Introduction

One of the most frequently discussed aspects of the special theory of relativity (STR) is the twin (or clock) paradox. The interest in this topic goes back to a comment that Einstein made in his original paper[5] about clocks in motion (Chapter 3. B). He concluded that a clock at rest on the Equator should have a slower rate that an identical clock at rest at one of the Earth's Poles. This led to speculation[136] that two twins would differ in age if one of them goes on a long journey into outer space at very high speed while the other remains at rest at his home base, specifically that the former would come back significantly younger than his stay-at-home brother. Subsequent experiments carried out on circumnavigating airplanes[42] proved definitively that at least Einstein's supposition about clock rate variations was correct. Whether that proves that the traveling twin would come back younger is clearly still a matter of great uncertainty, but the success of the experimentation with clocks raises another question that has thus far been largely ignored. *What happens to the size of the twin as he makes his way through space at such high speed?* The other bastion of STR is the FLC.[5] It states that he will decrease in size along the direction of motion but will otherwise stay the same, causing him to be distorted in much the same way as an image in a fun-house mirror.

Time and space are famously interconnected in relativity theory, so there is no reason for confining this discussion to the question of aging. As will be shown below, a clear picture emerges from both experiment and theory as to what happens to the dimensions of the twin as a result of his high-speed travels.

2. Changes in the Moving Twin's Circulation

One of the most basic characteristics about time dilation is that it is *uniform* for all clocks at rest in a given inertial system. This condition is required by the relativity principle (RP), since any violation of it would allow one to distinguish between one such rest frame and another. Furthermore, it also holds for all other types of natural timing devices, including the periods of biological processes occurring in plants and animals.

Consider, for example, the circulation of blood through the human body. If the accelerated twin A travels with constant speed $v = 0.866\ c$ relative to his initial location, his clocks, *i.e.* those that are at rest in his inertial system, will run exactly half as fast as those that are stationary in his brother B's rest frame. The RP therefore stipulates that A's blood will circulate with *twice* the period it would have under the same conditions if he was not moving relative to B.

There is another measure of circulation, however, namely the *speed* at which blood moves through the body. Relativity theory is quite clear on this point as well: the speed w between two fixed points is the *same* for all observers regardless of their state of motion. This principle has been applied in the discussion of muon decay,[43,44] for example, as well as in examples given in textbooks[137,138] to illustrate the effects of high-speed travel. It also can be said to be a consequence of the RP, since its violation would mean that w/c ratios would necessarily vary and therefore allow one to distinguish between different inertial systems on this basis (see Chapter 5.B).

The above comparisons raise an obvious question. How can twin B measure the period of A's blood circulation to be twice the normal value and still find that the speed with which the blood travels through A's veins and arteries is exactly the same as before his twin brother was rocketed into space? Moreover, how can one explain that the ratio of A's circulation period to the speed of the blood is independent of his orientation? Since speed is by definition the ratio of the distance D traveled to the elapsed time T required to complete the journey, there is only one acceptable answer for the above questions: *A's blood must travel twice as far during the circulation period under the conditions on the rocket ship, so that the $v = D/T$ ratio remains unchanged.* This conclusion in turn implies that twin A is now twice as tall and twice as wide as before he parted from twin B.

In summary, when one applies Einstein's two postulates of relativity directly to the problem of predicting changes in the dimensions of the accelerated twin A, a qualitatively different result is obtained than one expects from STR and its prediction of length contraction. Instead of his height decreasing by half along his direction of relative motion to twin B, it doubles instead. In addition, A's breadth does not remain unchanged but also doubles as a result of his changed state of motion. This is clearly uniform length expansion rather than the asymmetric length contraction claimed by STR and the FLC. As such, it is in perfect agreement with the conclusion of the clock riddle[32] discussed in Chapter 4. B, and also with the experimental findings for the transverse Doppler effect,[34,35,36] which show clearly that wavelengths increase uniformly in the rest frame of an accelerated light source in the same proportion as frequencies decline.

Appendix B

Proof of Eq. (67) Based on the VT ($\eta\eta' = \gamma^2$)

The VT in eq. (29) contains the function $\eta = (1 - vu_xc^{-2})^{-1} = (1 - vxt^{-1}c^{-2})^{-1}$, whereas its inverse in eq. (30) contains $\eta' = (1 + vu_x'c^{-2})^{-1} = (1 + vx't'^{-1}c^{-2})^{-1}$. In order for the Lorentz invariance condition of the ALT in eqs. (64-65) to be satisfied, it is necessary that the product of these two functions be equal to $\gamma^2 = (1\text{-}v^2c^{-2})^{-1}$. The proof is given below:

$$\eta\eta' = (1 - vxt^{-1}c^{-2})^{-1}(1 + vx't'^{-1}c^{-2})^{-1}$$
$$= 1/(1 - vxt^{-1}c^{-2})(1 + v\eta(x - vt)c^{-2}t^{-1})$$
$$= c^4t^2(c^2t - vx)^{-1}[c^2t + v\,(x - vt)(1 - vxc^{-2}t^{-1})^{-1}]$$
$$= c^4t^2(1 - vxc^{-2}t^{-1})[(c^2t - vx)(c^2t - v^2t)]^{-1}$$
$$= (c^4t^2 - vxc^2t)(c^4t^2 - c^2v^2t^2 - vxc^2t + v^3xt)^{-1}$$
$$= (c^4t^2 - vxc^2t)[(c^4t^2 - vxc^2t)(1 - v^2c^{-2})]^{-1}$$
$$= (1 - v^2c^{-2})^{-1} = \gamma^2.\ \text{QED}$$

Appendix C

Proof of the Scaling of Relative Velocities

The VT of eq. (29) relates the velocity components u_i and u_i' of an object from the standpoint of observers in two different rest frames (S and S′) that are moving with relative speed v along the x-axis of the coordinate system. The three equations for the respective x, y, z components are given below:

$$u_x' = (u_x - v)/(1 - vu_x c^{-2}) = \eta\,(u_x - v) \qquad \text{(C1a)}$$

$$u_y' = (1 - v^2c^{-2})^{0.5}(1 - vu_x c^{-2})^{-1}\,u_y = \eta\,\gamma^{-1}(v)\,u_y \qquad \text{(C1b)}$$

$$u_z' = (1 - v^2c^{-2})^{0.5}(1 - vu_x c^{-2})^{-1}\,u_z = \eta\,\gamma^{-1}(v)\,u_z, \qquad \text{(C1c)}$$

where $\gamma(v) = (1 - v^2/c^2)^{-0.5}$ and c is the speed of light in free space. Note the occurrence of the quantity $\eta = (1 - vu_x c^{-2})^{-1}$ in each case. It is a function of both the relative speed v of S and S′ as well as the parallel component u_x of the object's velocity relative to **v**.

The goal of the present discussion is to compare the *relative* velocity of two meta-stable particles observed in S and S′, respectively. To do this in a suitably general case, let us assume that particle P_1 moves along the x-axis with speed u_1 ($u_{1y} = u_{1z} = 0$) relative to an observer O at rest in S, while particle P_2 travels in a different direction with velocity components u_{2x} and u_{2y} relative to him ($u_{2z} = 0$). The first goal is to determine the relative velocity $\mathbf{u}_{21}$ of these two particles from O's perspective. To do this we need to use the relativistic VT of eqs. (C1a-c), and not simply compute the vector difference of the two velocities, that is, $\mathbf{u}_2 - \mathbf{u}_1$, as for the Galilean transformation. Accordingly, we find that (*i.e.*, with $\mathbf{v} = \mathbf{u}_1$ and $\mathbf{u} = \mathbf{u}_2$ in the VT):

$$u_{21x} = (u_{2x} - u_1)/\ (1 - u_{2x}u_1c^{-2}) \quad \text{(C2a)}$$

$$u_{21y} = u_{2y}\ /\gamma\ (u_1)\ (1 - u_{2x}u_1c^{-2}). \quad \text{(C2b)}$$

In effect, we are computing the velocity of P_2 from the vantage point of another observer (O″) *who is stationary in the rest frame of P_1*; O″ is therefore also moving with velocity $\mathbf{u_1}$ relative to O.

The next step is to compute the relative velocity of P_1 and P_2 ($\mathbf{u_{21}}'$) from the vantage point of observer O′ at rest in S′. We first need to compute the corresponding velocities of each particle relative to O′ based on their known velocity components relative to O. The VT allows us to do this:

$$u_{1x}' = u_1' = (u_1 - v)/(1 - vu_1c^{-2}) \quad \text{(C3a)}$$

$$u_{2x}' = (u_{2x} - v)/(1 - vu_{2x}c^{-2}) \quad \text{(C3b)}$$

$$u_{2y}' = u_{2y}\ /\gamma\ (v)(1 - vu_{2x}c^{-2}). \quad \text{(C3c)}$$

The final step is then to compute $\mathbf{u_{21}}'$ and compare the results with $\mathbf{u_{21}}$ in eqs. (C2a-b). In order to do this, we look upon O″ as an observer moving relative to O′. Using the VT and the velocity components from eqs. (C3a-c), the result is:

$$\begin{aligned}
u_{21x}' &= (u_{2x}' - u_1')\ /(1 - u_{2x}'u_1'c^{-2}) \\
&= (u_{2x} - v)/(1 - vu_{2x}c^{-2}) - (u_1 - v)/(1 - vu_1c^{-2}) \\
&\quad x[1 - (u_1 - v)\ (u_{2x} - v)c^{-2}(1 - vu_1c^{-2})(1 - vu_{2x}c^{-2})]^{-1} \\
&= c^2[(u_{2x} - v)(1 - vu_1c^{-2}) - (u_1 - v)(1 - vu_{2x}c^{-2})] \quad \text{(C4)} \\
&\quad x[c^2(1 - vu_1c^{-2})\ (1 - vu_{2x}c^{-2}) - u_1u_{2x} + vu_1 + vu_{2x} - v^2]^{-1} \\
&= c^2(u_{2x} - u_1)\ (1 - v^2c^{-2})/[c^2 - u_1u_{2x} - v^2 + v^2u_1u_{2x}c^{-2}] \\
&= (u_{2x} - u_1)/(1 - u_1u_{2x}c^{-2}) \\
&= u_{21x},
\end{aligned}$$

$$\begin{aligned}
u_{21y}' &= u_{2y}'/\gamma(u_1')(1 - u_{2x}'u_1'c^{-2}) \\
&= c^2u_{2y}(1 - v^2c^{-2})^{0.5}[1 - (u_1 - v)^2c^{-2}(1 - vu_1c^{-2})^2]^{0.5} \quad \text{(C5)} \\
&\quad x(1 - u_{2x}vc^{-2})^{-1}[c^2 - (u_1 - v)(u_{2x} - v)/(1 - vu_1c^{-2})(1 - vu_{2x}c^{-2})]^{-1}
\end{aligned}$$

$$= c^2 u_{2y}(1 - v^2c^{-2})^{0.5}[(c^2 + v^2u_1{}^2c^{-2} - u_1{}^2 - v^2)/$$

$$(c^2 - 2vu_1 + v^2u_1{}^2c^{-2})]^{0.5}$$

$$x[(c^2 - u_1u_{2x})(1 - v^2c^{-2})/(1 - vu_1c^{-2})]^{-1}$$

$$= c^2u_{2y}\ [(c^2 - u_1{}^2)c^{-2}]^{0.5}/(c^2 - u_1u_{2x})$$

$$= c^2u_{2y}/\gamma\ (u_1)(c^2 - u_1u_{2x}) = u_{2y}/\gamma(u_1)(1 - u_1u_{2x}c^{-2})$$

$$= u_{21y}.$$

It is thus seen that O and O′ agree on both the direction and the speed with which the two rest frames are moving relative to one another, as was to be shown. There is no difference between the Galilean velocity transformation and Einstein's relativistic version[5] in this respect.

Appendix D

Uniform Scaling of Electromagnetic Quantities

1. Introduction

It has been shown in Chapter 7. C that one can conveniently describe relativistic effects such as time and mass dilation by assuming that the units of physical quantities vary in a well defined manner depending on the state of motion of the observer and the object of measurement. A similar conclusion has been reached for gravitational interactions.[56] The basic idea is simply to look upon the slowing down of clocks upon acceleration as a change in the unit of time in their current rest system.[54] In effect, there is a *uniform scaling* relationship between the values of a given physical quantity for two observers who are in relative motion to another and/or are at different positions in a gravitational field. In kinematics there is an important simplifying feature in such a discussion because of the fact that all physical quantities can be expressed in terms of the units of length (m), time (s) and mass (kg). It is therefore sufficient to know how the latter three quantities scale in order to deduce the corresponding relationships for other properties such as angular momentum, energy and force. The key point is that knowledge of how the meter, second and kilogram vary from one of state of motion to another allows one to apply this information to the scaling of all other kinematic quantities.

The situation is not so simple when it comes to the scaling of electromagnetic quantities such as charge and electric and magnetic fields, however. The fundamental equations in this field of physics are expressed in terms of a rather large set of quantities that are not directly connected to the above three units of kinematics. The Giorgi system was introduced[139] in 1901 in order to ensure that the results of electromagnetic calculations ultimately can be expressed in terms of the above (mks) system of units, however. The question that arises in the present context is how do electromagnetic quantities such as charge, magnetic induction, electric displacement and inductance, to name just a few, vary with the state of motion of the observer or object of the measurement as well as with their respective locations in a gravitational field. In order to answer this in a definitive manner, it proves helpful to carry out an extensive review of the Giorgi system of electromagnetic units and particularly to carefully understand how it is related through experiment and theory to the mks system.

2. Coulomb's Law and the Definition of Electric Charge

The simplest way to begin this analysis is to consider how Coulomb's Law is formulated in the Giorgi system. The force $\mathbf{F}_e$ in Newtons (1 N = 1 kg m/s^2) between two electric charges q_i and q_j (expressed in Coul) separated by a distance of $\mathbf{r}_{ij}$ m is given by the vector relation:

$$\mathbf{F}_e = q_i q_j \mathbf{r}_{ij}/4\pi\varepsilon_0 r_{ij}^3, \tag{D1}$$

where ε_0 is the permittivity of free space. Note that the units for ε_0 are given in such a way (Coul^2/Nm^2) so as to insure that the final result is expressed in the mks unit of force (N). The point that needs to be emphasized with regard to this equation is that it serves as a definition of both electric charge and ε_0. In order to satisfy the above requirement in the mks system, it is actually only necessary that the unit for the product of two electric charges q_i q_j divided by ε_0 is Nm^2. This shows that there is an *inherent redundancy* in any system of electromagnetic units *that cannot be removed by experiment.*

We can take advantage of this situation in the context of a uniform scaling procedure by defining the unit of electric charge to be some combination of mks units, that is, without introducing a new unit such as the Coulomb for this purpose. We just have to make a corresponding choice of unit for ε_0 to ensure that the force $\mathbf{F}_e$ in eq. (D1) is expressed in N. For example, the unit of electric charge could be defined to be the same as for energy (1 J = 1 N m). That would simply mean that the unit of permittivity is N, since then $q_i q_j/\varepsilon_0$ in eq. (D1) has the required unit of Nm^2 mentioned above. Since we know how N and m scale with the state of motion and position in a gravitational field,[56] we can immediately determine how electric charge and ε_0 scale on this basis as well. The very arbitrariness of the above choice of units might tend to make one feel skeptical about such a procedure. What it actually shows, however, is that such quantities are only defined *indirectly* by experiment. As much as we have gotten used to the idea of electric charge over time, it should not be forgotten that there is no other way to determine its magnitude experimentally than to measure the force exerted between it and another charge when they are a certain distance apart.

It is no less permissible to choose a system of electromagnetic units such that ε_0 is dimensionless. This is in fact what is done with the older Gaussian set of units in which charge is expressed in esu. In that system the quantity $4\pi\varepsilon_0$ in Coulomb's Law is missing entirely. One can do this and still remain in the mks system by defining the unit of electric charge to be $N^{0.5}m$. Again, there is no *a priori* reason for avoiding such a choice because charge is only defined experimentally through eq. (D1).

There is only one other relationship that must be satisfied in order to extend such an mks-type system to the description of magnetic interactions. The constant μ_0 in the law of Biot and Savart[140] must satisfy the equation from Maxwell's electromagnetic theory:

$$\varepsilon_0 \mu_0 c^2 = 1, \tag{D2}$$

where c is the speed of light in free space. The unit in the Giorgi system is N/Amp^2 or $Ns^2/Coul^2$. If the unit of ε_0 is N, it follows from eq. (D2) that the corresponding unit for μ_0 is s^2/Nm^2. Alternatively, if ε_0 is to be dimensionless, then the unit for μ_0 becomes s^2/m^2. That means in turn that in the latter system of units, the values of all three quantities in eq. (D2) would be the same for all observers independent of their relative speed to one another (provided that they were all located at the same gravitational potential[56]).

Once the unit of electric charge has been fixed in the mks system, the corresponding units for all other quantities that occur in

Table D1. Correlation of the units of electromagnetic quantities in various systems. The standard Giorgi system is compared with two alternatives, the Nms and $N^{0.5}$ms systems, whose units are exclusively multiples of N, m and s in the standard mks system for strictly mechanical variables. The quantities are also subdivided into K-type scaling classes, as discussed in Sect. D3.

Quantity	**Symbol**	**Giorgi**	**Nms**	**$N^{0.5}$ms**	
Electric charge	q	Coul	Nm	$N^{0.5}$m	K
Permittivity	ε or $ε_0$	$Coul^2/Nm^2$	N	____	K^2
Current/mmf	I	Amp	Nm/s	$N^{0.5}$m/s	K
Permeability	μ or $μ_0$	N/Amp^2	s^2/Nm^2	s^2/m^2	K^{-2}
Potential/emf	V	Volt	____	$N^{0.5}$	K^{-1}
Resistance/impedance	R/Z	Ohm	s/Nm	s/m	K^{-2}
Electric field	**E**	Volt/m	1/m	$N^{0.5}$/m	K^{-1}
Volume charge density	ρ	$Coul/m^3$	N/m^2	$N^{0.5}/m^2$	K
Surface charge density	σ	$Coul/m^2$	N/m	$N^{0.5}$/m	K
Electric dipole moment	$\boldsymbol{\mu}_e$	mCoul	Nm^2	$N^{0.5}m^2$	K
Electric quadrupole moment	Q_{ij}	m^2 Coul	Nm^3	$N^{0.5}m^3$	K
Electric polarization	**P**	$Coul/m^2$	N/m	$N^{0.5}$/m	K
Electric displacement	**D**	$Coul/m^2$	N/m	$N^{0.5}$/m	K
Electric susceptibility	χ	Coul/mVolt	N	____	K^2
Polarizability	α	m^2Coul/Volt	Nm^3	m^3	K^2
Coefficient of potential	p_{ij}	Volt/Coul	1/Nm	1/m	K^{-2}
Capacitance/coeff. of capacitance	C or c_{ij}	Coul/Volt	Nm	m	K^2
Current density	**J**	$Coul/m^2s$	N/ms	$N^{0.5}$/ms	K
Conductivity	g	Coul/msVolt	N/s	1/s	K^2
Resistivity	η	msVolt/Coul	s/N	s	K^{-2}
Magnetic flux	Φ	Weber	s	$N^{0.5}$s	K^{-1}
Magnetic induction/field	**B**	$Weber/m^2$	s/m^2	$N^{0.5}s/m^2$	K^{-1}
Magnetic vector potential	**A**	Weber/m	s/m	$N^{0.5}$s/m	K^{-1}
Magnetic scalar potential	U*	Amp	N/ms	$N^{0.5}$m/s	K
Magnetic dipole moment	**M**	m^2Amp	Nm^3s	$N^{0.5}m^3/s$	K
Magnetization	**M**	Amp/m	N/s	$N^{0.5}$/s	K
Inductance	L	Henry	s^2/Nm	s^2/m	K^{-2}
Magnetic current per unit area	$\mathbf{J}_m$	Amp/m^2	N/ms	$N^{0.5}$/ms	K
Magnetic intensity	**H**	Amp/m	N/s	$N^{0.5}$/s	K
Reluctance	*R*	Amp/Weber	Nm/s^2	m/s^2	K^2
Admittance	Y	Mho	Nm/s	m/s	K^2

the theory of electricity and magnetism are determined by the standard equations in which they occur. A fairly extensive list of such quantities illustrating this point is given in Table D1. The corresponding units are always given in terms of those of force, length and time in the mks system. Two sets are given in each case, one in which the unit of charge is Nm and the other in which it is $N^{0.5}$m. The former is referred to as the Nms system so as to distinguish it from the standard mks system for purely kinematic quantities, the other as the $N^{0.5}$ms system, in which ε_0 is dimensionless.

Just a few examples will be given below which emphasize the practicality of the concepts introduced above. The unit of potential/emf U is dimensionless in the Nms system since it is proportional to electric charge and inversely proportional to ε_0 and a distance given in m. It has the unit of $N^{0.5}$ in the other system based on the same definition. Since the electric field **E** is the gradient of a potential, it follows that it has a unit of m^{-1} in the Nms system and $N^{0.5}$/m in the other. The unit of current I is Nm/s in the former case, while that of resistance R (I = V/R) is accordingly s/Nm. In the $N^{0.5}$ms system, R has the unit of s/m, *i.e.* the reciprocal of that of velocity, whereas the unit for I is $N^{0.5}$m/s.

In the Giorgi system of units, the magnetic force $\mathbf{F_m}$ for a given charge q moving with velocity **v** in magnetic field **B** is defined as:

$$\mathbf{F_m} = q\,\mathbf{v} \times \mathbf{B}. \tag{D3}$$

It therefore follows that **B** has the unit of s/m^2 in the Nms system and $N^{0.5}s/m^2$ in the $N^{0.5}$ms system. The Nms unit of magnetic flux (Weber or m^2Tesla in the Giorgi system) is s, consistent with the requirement that an induced emf, which is dimensionless in the Nms system of units, is given by the derivative of the magnetic flux with respect to time. In the $N^{0.5}$ms system its unit is $N^{0.5}$s. It is easy to show that the units are consistent for Maxwell's equations in both of these systems of units. For example, the differential form of Faraday's law of electromagnetic induction,

$$\mathbf{curl\ E} = -\,\partial \mathbf{B}/\partial\, t, \tag{D4}$$

has the units of m^{-2} on both sides in the Nms system and $N^{0.5}m^{-2}$ in the other.

3. A Simple Scaling Procedure for Electromagnetic Quantities

The interdependency of the definitions of electric charge q and permittivity ε_0 also presents other options for the choice of units for electromagnetic quantities than those of the Giorgi system. The esu system of units[141] employs a much smaller unit of electric charge than Coul, for example, which therefore makes it unnecessary to include the $4\pi\varepsilon_0$ factor in eq. (1), which is to say that in this system of units, $\varepsilon_0 = 1/4\pi$. The system of atomic units, in which the electronic charge e serves as the unit of electric charge, makes the same choice for ε_0. In the present section we will illustrate how the various electromagnetic units of the Giorgi system can be modified in a systematic manner so that the latter condition is also fulfilled for mks units.

To begin this discussion it is important to note that the value of ε_0 in the Giorgi system is based directly on the speed of light in mks units: the value of $4\pi\varepsilon_0$ is equal to 10^7c^{-2}. Since the speed of light in free space is no longer measured but is simply defined by international convention to have the value of 2.99792458×10^8 ms^{-1}, it follows that there is also no need to determine quantities such as the Coulomb (Coul) and ε_0 that are ultimately based on the value of c. A convenient quantity with which to scale the various standard Giorgi units is $K = (4\pi\varepsilon_0)^{-0.5} = 10^{-3.5}c = 94803$. In the following we will refer to the new set of units as the KNms system. First, we define the corresponding value of the permittivity as $\varepsilon_0' = K^2\varepsilon_0$, so that $4\pi\ \varepsilon_0' = 1$ N. In general, the units in the new system are those given in Table D1 under the Nms heading, that is, with the unit of electric charge equal to 1 J = 1 Nm. It should be clear, however, that the numerical value attached to ε_0' in the new system is completely independent of this choice. One could just as well choose the unit of charge to be $N^{0.5}m$, for example, or any other combination of N, m and s, as long as one adheres to the requirements already discussed in Appendix D.2.

The objective in changing the numerical values of electromagnetic constants such as ε_0 is clearly to simplify computations in this important area of physics. One of the problems with changing over from the Giorgi to the Gaussian system of units is that in many cases this requires using different formulas for the same interaction. One can avoid this difficulty by agreeing at the outset that all formulas in the new KNms system will be the same as for the Giorgi system, since the latter have become standard over the past century. Let us consider eq. (D1) as the first example. In order to retain the same form for this equation while using the above value for ε_0', it is simply necessary to change the numerical value of each electric charge. Specifically, one has to change the unit of charge to K^{-1} Coul. This means that the value of the electronic charge (e′) becomes K times larger than the standard value in Coul, *i.e.* $e' = 94803 \times 1.602 \times 10^{-19}$ J $= 1.5187 \times 10^{-14}$ J. In effect then, the change from the Giorgi to the KNms system of units occurs by multiplying both the numerator and denominator in eq. (D1) by the same factor (K^2). The result is that one has the same form for eq. (D1) as in the Gaussian or atomic unit versions, *i.e.* where $4\pi\varepsilon_0 = 1$ and thus does not appear explicitly.

The main point that the above discussion reveals is that it is useful to divide the variables that commonly occur in the theory of electricity and magnetism into classes according to the way in which their numerical values need to be scaled. In the KNms system, this means that each such variable simply needs to be associated with a specific power of K. This information has also been given in Table D1 in each case. Since $\varepsilon_0' = K^2\varepsilon_0$, for example, it is necessary to multiply the Giorgi value for μ_0 by K^{-2} in order to be consistent with eq. (D2), that is, without changing the value of c. As a result, $\mu_0' = 4\pi c^{-2}$. Again, the preferred approach is not to eliminate ε_0' and μ_0' from the formulas in the KNms system, rather only to change their numerical values relative to those in the Giorgi mks system so that the form of the standard equations in the latter system is completely retained.

Other quantities that belong to the same K-class in Table D1 as electric charge are charge densities ϱ and σ, dipole moment $\boldsymbol{\mu}$, quadrupole moment Q, current I, current density **J**, magnetic dipole moment **m**, magnetization **M** and magnetic intensity **H**. The corresponding quantities of K^{-1} type are: electric potential U or emf, electric field **E**, magnetic field (or induction) **B**, magnetic flux Φ and magnetic vector potential **A**. A check of all formulas in which the latter quantities appear shows that they always occur with counterparts in the K class mentioned first, as, for example, q and **B** in eq. (D3) or q and **E** in the corresponding expression for electric force.

Table D2. Conversion of various electromagnetic units from the Giorgi to the KNms system discussed in Sect. D3.

Quantity	Giorgi	KNms
Electric charge	1 Coul	$10^{-3.5}c$ Nm
Electric current	1 Amp	$10^{-3.5}c$ Nm/s
$4\pi\varepsilon_0$	$10^7/c^2$ Coul^2/Nm^2	1 N
$\mu_0/4\pi$	10^{-7} N/Amp^2	$1/c^2 s^2/\text{Nm}^2$
Electric field	1 Volt/m	$10^{3.5}/c\ m^{-1}$
Potential	1 Volt	$10^{3.5}/c$
Magnetic induction	1 Weber/m^2	$10^{3.5}/c\ s/m^2$
Magnetic intensity	1 Amp/m	$10^{-3.5}c$ N/s
Magnetic flux	1 Weber	$10^{3.5}/c$ s
Electric displacement/polarization	1 Coul/m^2	$10^{-3.5}c$ N/m
Capacitance	1 Farad = Coul/Volt	$10^{-7}c^2$ Nm
Inductance	1 Henry	$10^7/c^2\ s^2/\text{Nm}$

Some quantities do no not have to be scaled at all (K^0-type). They include all dimensionless quantities such as magnetic susceptibilities and refractive indices. The same is of course true for all non-electromagnetic quantities such as force, energy and angular momentum. A less trivial example is the Poynting vector (**E** × **H**), which is a product of a K^{-1}- and K-type variable, respectively. All other commonly occurring quantities are either of K^2- or K^{-2}-type. In addition to ε_0 among the former are the dielectric constant ε and electrical susceptibility χ (Table D1), as well as polarizability, capacitance, reluctance, conductivity and admittance. Some examples of K^{-2}-type are in addition to μ_0: permeability μ, resistance, coefficient of potential p_{ij}, resistivity η and inductance L. The latter quantity is defined as $d\Phi/dI$, which is a ratio of a K^{-1}–type quantity to the current, which is of K-type.

The conversion factors between the Giorgi and the present KNms systems of electromagnetic units for a number of the most commonly used quantities are given in Table D2. Unlike the case for the corresponding conversion between the Gaussian and

Giorgi systems,[141] the formulas in which they are to be used respectively are exactly the same, as discussed above. To be specific, we have given these factors as functions of c rather than of K itself. Clearly, any other value of K could be used while still allowing the Giorgi formulas to be retained in the new system of units. The value of the electric charge in any such system of units is K times that of the numerical value in the Giorgi system ($e = 1.602 \times 10^{-19}$). As long as one adheres to the scheme of dividing the variables into K-type classes according to the prescriptions of Table D1, this information is sufficient to characterize any new system of this type. In other words, the scaling procedure is always perfectly defined by the value chosen for K in a specific instance.

4. Conclusion

The fact that different sets of kinetic and gravitational scaling factors for quantities such as electric charge and magnetic fields will lead to the same predictions emphasizes that there is a qualitative difference between mechanical quantities such as distances and lifetimes and their counterparts in the field of electromagnetism. Over time we have come to believe that we experience electric charge directly, for example, but the fact is that this is only a theoretical quantity that allows us to calculate the Coulomb force between particles at any distance. Force is something very tangible whereas charge is not, despite what our intuition and training tell us. The determination of a **B** field also can only be accomplished by measuring the force between electric charges in relative motion. All the other myriad quantities mentioned in Table D1 have definitions that in one way or another involve explicit measurements of such forces.

It should therefore not be surprising to discover that there is an element of arbitrariness in the way the kinetic and gravitational scaling of electromagnetic quantities can be specified theoretically. As long as one remains consistent, that is, adheres to the simple restrictions outlined in Appendix D.2, one system of electromagnetic units will inevitably lead to the same set of measured values for force, distance and elapsed time as any other. This realization is at odds with what one commonly finds in the literature,[142] namely that electric charge is a relativistic invariant. The argument given for the latter conclusion is typically based on the RP and the fact that *in situ* measurements of a given charge are completely independent of the state of motion of the observer. The same can be said of the lifetimes of meta-stable particles, however, and we know that these quantities do change with the state of motion or position in a gravitational field.

The discovery of time dilation actually requires that one be more specific in defining the RP than is usually the case (see Chapter 6.C). One should add that, although the laws of physics are the same in all inertial systems, the *units* in which the various quantities contained in them are expressed are *not* necessarily equal. *In situ* measurements do not vary from one inertial system to another for the simple reason that the scaling of such units is *completely uniform*. The same holds true for *in situ* measurements at different gravitational potentials. The only way that one can discern how such units change is to carry out measurements in which the observer is not at the same gravitational potential as the object or is in relative motion to it. Even when this is done, however, one is still free to choose how electromagnetic quantities should scale because of the redundancy in the definition of electric charge and permittivity discussed in Appendix D.2.

Appendix E

Objectivity of Measurement

The scientific theory of measurement was born out of the needs of commerce to establish fair and just trading practices. Buyers and sellers could disagree about the quality of merchandise but it was apparent that once the bargaining process had been completed, there had to be quantitative standards for lengths and weights on which to base the final price for a given sale. It was the nature of the process that one could not expect businessmen from different backgrounds to agree on what those standards were, but at least they could agree instinctively on the simple question of which quantity of a given material was larger or smaller. This innate belief in the objectivity of the measurement process was all but taken for granted over a period of at least several millennia until something extraordinary occurred. The study of electricity and magnetism led to the conclusion that space and time are inextricably mixed and that two observers in relative motion can therefore disagree as to which of two clocks runs slower and which of two distances is longer or shorter. As a result, it became the common belief of the physics community that the measurement process is *subjective* and depends on the perspective of the observer, despite the aforementioned longstanding experience to the contrary.

Galileo played a key role in many of these developments by virtue of his introduction of the Relativity Principle (RP) in 1632. It was in reaction to his findings and those of his contemporaries that the earth travels at a very high speed while orbiting the sun. Using the analogy of a ship traveling on perfectly calms seas, he argued that it would be impossible for passengers located below deck to distinguish whether they were in motion or were still located at the port of embarkation. Einstein used this as one of his two postulates in order to derive the Lorentz transformation (LT), the other being the constancy of the speed of light in free space independent of the motion of the source or the observer. Rejecting the essential objectivity of the measurement process was by far not the only revolutionary aspect of Einstein's Special Theory of Relativity (STR). He agreed with Poincaré that there was no reason to exclude the possibility of remote non-simultaneity of events for different observers. This conclusion followed directly from the space-time mixing characteristic of the LT, which also left open the possibility of time reversal, *i.e.* the possibility of two observers disagreeing on the time-order of events. Another point of agreement with Poincaré was the existence of group properties of successive Lorentz transformations, which characteristic played an important role in the latter's independent derivation of these equations. Einstein's new theory also incorporated the length-contraction phenomenon postulated independently by FitzGerald and Lorentz (FLC), although it was no longer based directly on the results of the Michelson-Morley experiment. The amount of the contraction of objects located in a different inertial frame varied with their orientation, creating the image of their changing shape upon acceleration relative to the observer. The phenomenon of time dilation also follows directly from the LT, and both effects are expected to bear a symmetric relationship for two observers in relative motion, consistent with the overall subjectivity of STR. However, Einstein also concluded in his paper that one should expect clocks located at the Equator to run slower than their identical counterparts at either of the earth's Poles by virtue of the greater acceleration of clocks located in the former position.

The LT carried with it a new symmetry condition known as Lorentz invariance. It is widely considered to be the inescapable consequence of Einstein's two postulates of relativity. This relationship put restrictions on the nature of physical laws that ultimately had a profound effect on developments in the theory of elementary particles and high-energy interactions. A key aspect of STR is that laws of physics in different inertial systems must be related through a Lorentz transformation of their respective space-time coordinates. This requirement led to the introduction of the word "covariance" into the physics literature. Energy and momentum are related by a similar transformation, which in turn leads to the mass-energy equivalence relation and Einstein's most widely publicized equation, $E = mc^2$.

Perhaps the greatest benefit of STR has been to suggest new experimental investigations to test its validity. A great majority of these have involved the phenomenon of time dilation. This work has led to many technical advances, including the Global Positioning System (GPS) for measuring distances on the earth's surface that has become a staple of modern life. The first successful experiment of this type was carried out by Ives and Stilwell in 1938 confirming Einstein's prediction of a *transverse* Doppler effect. Since that time there have been experiments with decaying elementary particles, the Mössbauer effect and atomic clocks that have greatly improved our understanding of time dilation. However, the symmetry principle expected from the LT has never been verified in these studies. Instead, it has been found that it is always possible to know which of two clocks runs slower. The indication is therefore that the subjectivity of measurement predicted by the STR does not actually occur in nature.

For this reason, a goal of the present book is to critically examine the arguments that Einstein used in his original work[5] to derive the LT and to understand how the objectivity of the measuring process can be reconciled with Einstein's two postulates of relativity. The surprising result is that Einstein made an undeclared assumption in his derivation, a "hidden" third postulate, according to which a certain function could only depend on the value of the relative speed of the two inertial systems considered in the transformation. This assumption is then shown to be incorrect because it leads to a self-contradiction in STR. The FLC predicts the opposite behavior as when the time-dilation effect is used in combination with the light-speed postulate to measure distances (clock riddle[32]). The contradiction can be removed by making a different assumption from Einstein's that takes into account experimental data for time dilation which were not available in 1905. The rates of proper clocks are assumed to follow a strict proportionality law between inertial systems. The proportionality constant does not depend on the locations of the clocks but rather only on their respective states of motion (as with STR, gravitational effects on these rates are not considered in this development). The new assumption is still consistent with Einstein's two postulates of relativity and with the relativistic velocity transformation (VT). As a result, all known experimental verifications of Einstein's original theory are left intact, such as the aberration of starlight at the zenith and the Fresnel light-drag experiment, as well as all the aforementioned results for the transverse Doppler effect and the lifetimes of meta-stable particles.

Nonetheless, adopting this new assumption is not without significant consequences for Einstein's original theory. The LT is eliminated, along with Lorentz invariance, the FLC and space-time mixing. There is no longer any need to assume that remote non-simultaneity and time reversal of events can occur, and the VT takes on a significantly more prominent role in the resulting version of relativity theory. The term "covariance" loses much of its significance. Emphasis is placed instead on the way the physical properties of objects vary with their state of motion. A key goal becomes the quantitative prediction of the extent of such changes for all types of physical properties, including those defined for electromagnetic interactions. Because the theory subscribes to a perfectly objective description of the measurement process, it is possible to ascribe a set of standard units to each inertial system. To a co-moving observer these units do not appear to change with his state of motion, consistent with Galileo's RP. This is because all such changes are perfectly uniform within a given rest frame. This state of affairs leads to an amended version of the RP: The laws of physics are the same in all inertial systems, *but the units in which they are expressed differ with the state of motion of the observer.*

The physical properties of objects also vary with their position in a gravitational field.[56] A constant defined in Einstein's landmark paper of 1907[38] in which he enunciated the Equivalence Principle (EP) between kinematic and gravitational acceleration turns out to play a similar role as that for time dilation in describing the magnitudes of such changes. The concept of gravitational scaling of units allows one to develop a simple alternative to General Relativity (GTR)[76]. With the aid of a method presented by Schiff in 1960[78], it is possible to obtain quantitative agreement with the three major predictions of the latter theory, namely the gravitational red shift of light frequencies, the angle of displacement of the images of stars during solar eclipses and the precession of the perihelion of planets orbiting the sun. The resulting theory differs from Newton's Universal Law of Gravitation primarily because of its recognition that the speed of light is a finite constant, one that nonetheless varies with the location of the source in a gravitational field.

Endnotes

[1] C. W. Kilmister, in *Special Theory of Relativity* (Pergamon Press, Oxford, 1970), p. 1.
[2] W. Voigt, *Goett. Nachr.*, 1887, p. 41.
[3] A. A. Michelson and E. W. Morley, *Am. J. Sci.* 34, 333 (1887); L. Essen, *Nature* 175, 793 (1955).
[4] A. Pais, '*Subtle is the Lord…' The Science and Life of Albert Einstein* (Oxford University Press, Oxford, 1982), pp. 121-134.
[5] A. Einstein, *Ann. Physik* 322 (10), 891 (1905).
[6] R. D. Sard, in *Relativistic Mechanics* (W. A. Benjamin, New York, 1970), p. 59.
[7] H. A. Lorentz, *Versl. K. Ak. Amsterdam* 10, 793 (1902); *Collected Papers*, Vol. 5, p. 139.
[8] H. Poincaré, *C. R. Ac. Sci. Paris* 140, 1504 (1905); *Oeuvres*, Vol. 9, p. 489.
[9] H. Goldstein, in *Classical Mechanics* (Addison-Wesley, Reading, 1950), p. 189.
[10] G. F. FitzGerald, *Science* 13, 390 (1889).
[11] H. A. Lorentz, *Versl. K. Ak. Amsterdam* 1, 74 (1892); *Collected Papers*, Vol. 4, p. 219.
[12] A. Pais, '*Subtle is the Lord…' The Science and Life of Albert Einstein* (Oxford University Press, Oxford, 1982), p. 144.
[13] R. T. Weidner and R. L. Sells, *Elementary Modern Physics* (Allyn and Bacon, Boston, 1962), p. 33.
[14] M. von Laue, *Relativitätstheorie*, Vol. 1, 7th Ed. (Vieweg, Braunschweig, 1955), p. 36.
[15] C. M. Will, *Phys. Rev.* D 45, 403 (1992).
[16] M. von Laue, *Ann. Physik* 23, 989 (1907).
[17] J. D. Jackson, Classical Electrodynamics (J. Wiley and Sons, New York, 1962), pp. 362-363.
[18] Lord Rayleigh, *The Theory of Sound*. 2nd Ed. (Macmillan, London, 1894-6; Dover Publications, New York, 1945).
[19] L. Brillouin, *Wave Propagation and Group Velocity* (Academic Press, New York, 1960), p. 1.
[20] A. A. Michelson, *Rep. Brit. Assoc. Montreal*, 1884, p. 56
[21] R. Wang, Y. Zheng, A. Yao and D. Langley, *Phys. Lett. A* 312, 7 (2003); R. Wang, Y. Zheng and A. Yao, *Phys. Rev. Lett.* 93, 143901 (2004); R. Wang, *Galilean Electrodynamics* 16, 23 (2005).
[22] G. Sagnac, *C. R. Acad. Sci. Paris* 157, 708 (1913).
[23] T. E. Phipps, Jr., *Old Physics for New* (Apeiron, Montreal, 2006), p. 42.
[24] R. J. Buenker, *Apeiron* 20, 27 (2013)
[25] R. D. Sard, in *Relativistic Mechanics* (W. A. Benjamin, New York, 1970), p. 105.
[26] R. D. Sard, in *Relativistic Mechanics* (W. A. Benjamin, New York, 1970), p. 285-290.
[27] L. H. Thomas, *Nature* 117, 514 (1926); *Phil. Mag.* 3, 1 (1927).
[28] J. D. Jackson, *Classical Electrodynamics* (J. Wiley and Sons, New York, 1962), p. 367.
[29] L.I. Schiff, *Proc. Natl. Acad. Sci.* U. S. 46, 871 (1960).
[30] H. Dingle, *Science at the Crossroads* (Martin, Brian and O'Keefe, London, 1972).
[31] T. E. Phipps, Jr., *Old Physics for New* (Apeiron, Montreal, 2006), pp. 123-164.
[32] R. J. Buenker, *Apeiron* 19, 84 (2012).
[33] R. J. Buenker, *Apeiron* 15, 254 (2008).
[34] H. E. Ives and G. R. Stilwell, *J. Opt. Soc. Am.* 28, 215 (1938); 31, 369 (1941).
[35] G. Otting, *Physik. Z.* 40, 681 (1939).
[36] H. I. Mandleberg, L. Witten, *J. Opt. Soc. Am.* 52, 529 (1962).
[37] H. J. Hay, J. P. Schiffer, T. E. Cranshaw and P. A. Egelstaff, *Phys. Rev.* Letters 4, 165 (1960).

[38] A. Einstein, *Jahrb. Rad.Elektr.* 4, 411(1907).
[39] W. Kuendig, *Phys. Rev.* 129, 2371 (1963).
[40] D. C. Champeney, G. R. Isaak,and A. M. Khan, *Nature* 198, 1186 (1963).
[41] W. Sherwin, *Phys. Rev.* 120, 17 (1960).
[42] J. C. Hafele and R. E. Keating, Science 177, 166 (1972); J. C. Hafele and R. E. Keating, *Science* 177, 168 (1972).
[43] B. Rossi, K. Greisen, J. C. Stearns, D. K. Froman, P. G. Koontz, *Physical Review* 61,675 (1942).
[44] D. S. Ayres, D. O. Caldwell, A. J. Greenberg, R. W. Kenney, R. J. Kurz, B. F. Stearns, *Physical Review* 157, 1288 (1967); A. J. Greenberg, Thesis, Berkeley (1969).
[45] W. Rindler, in *Essential Relativity*, Springer-Verlag, New York, 1977, p. 44.
[46] T. Wilkie, *New Scientist* 27, 258 (1983).
[47] R. J. Buenker, *Apeiron* 17, 99 (2010).
[48] A. Laub, T. Doderer, S. G. Lachenmann, R. P. Huebner and V.A. Oboznov, *Phys. Rev.* Letters 75, 1372 (1995).
[49] L. de Broglie, *Compt. Ren.* 177, 507 (1923).
[50] en.wikipedia.org/wiki/Length_contraction.
[51] R. J. Kennedy and E. M. Thorndike, *Phys. Rev.* 42, 401 (1932).
[52] R. J. Buenker, *Apeiron* 16, 96 (2009).
[53] R. J. Buenker, *Apeiron* 19, 282 (2012).
[54] R. D. Sard, in *Relativistic Mechanics* (W. A. Benjamin, New York, 1970), p. 95.
[55] R. J. Buenker, *Apeiron* 19, 218 (2012).
[56] R. J. Buenker, *Apeiron* 15, 382 (2008).
[57] R. D. Sard, in *Relativistic Mechanics* (W. A. Benjamin, New York, 1970), p. 364.
[58] W. Rindler, in *Essential Relativity* (Springer-Verlag, New York, 1977), p. 55.
[59] C. M. Will, in *Was Einstein Right?: Putting General Relativity to the Test.* 2nd ed. (Basic Books Inc; New York 1993), p. 272.
[60] OPERA collaboration, T. Adam *et al.*, "Measurement of the neutrino velocity with the OPERA detector in the CNGS beam", arXiv:1109.4897 [hep-ex].
[61] A. Sommerfeld, *Z. Phyzik* 8, 841 (1907).
[62] A. M. Steinberg, P. G. Kwiat and R. Y. Chiao, *Phys. Rev.* Lett. 71, 708 (1993).
[63] A. Enders and G. Nimtz, J. Phys. I (France) 3, 1089 (1993); G. Nimtz, *Phyzik in unserere Zeit* 28, 214 (1997).
[64] Y. Aharonov and L. Vaidman, *Phys. Rev.* A 41, 11 (1990).
[65] R. D. Sard, in *Relativistic Mechanics* (W. A. Benjamin, New York, 1970), p. 138.
[66] A. H. Bucherer, *Phys. Zeit.* 9, 755 (1908).
[67] R. D. Sard, in *Relativistic Mechanics* (W. A. Benjamin, New York, 1970), p. 118.
[68] G. N. Lewis and R. Tolman, *Phil. Mag.* 18, 510 (1909).
[69] R. D. Sard, in *Relativistic Mechanics* (W. A. Benjamin, New York, 1970), p. 142.
[70] R. D. Sard, in *Relativistic Mechanics* (W. A. Benjamin, New York, 1970), p. 356.
[71] H. A. Lorentz, in Versuch Einer Theorie der Electrischen und Optischen Erscheinungen in Bewegten Körpern, *Collected Papers*, Vol. 5 (Brill, Leiden, 1895), p. 1.
[72] R. J. Buenker, "Expressing the Units of Electricity and Magnetism Directly in the mks System," accepted for publication in the *J. Chem. Phys.* (Russia), 2013.
[73] J. G. von Soldner, *Berliner Astr. Jahrb.*, pp. 161-172 (1804).
[74] A. Pais, *'Subtle is the Lord…' The Science and Life of Albert Einstein* (Oxford University Press, Oxford, 1982), p. 200.
[75] A. Einstein, *Ann. Physik* 35, 898 (1911).
[76] A. Einstein, *Ann. Physik* 49, 769 (1916).
[77] F. W. Dyson, A. S. Eddington, and C. Davidson, *Phil. Trans. Roy. Soc.* (London) 220A, 291 (1920).
[78] L. I. Schiff, Am. J. Phys. 28, 340 (1960).

[79] R. J. Buenker, *Apeiron* 15, 509 (2008).
[80] S. Weinberg, *Gravitation and Cosmology* (Wiley, New York, 1972), p. 198.
[81] R. H. Dicke, *The Theoretical Significance of Experimental Relativity*, (Gordon and Breach, New York, 1964), p. 27.
[82] W. W. Campbell, and R. J. Trumpler, *Lick Observ. Bull.* XIII, 130-160 (1928).
[83] R. J. Buenker, *Apeiron* 15, 338 (2008).
[84] R. J. Buenker, *J. Chem. Phys.* (Russia) 24, No. 4, 29 (2005).
[85] R. A. Houston, *Proc. Roy. Soc. Edinburgh* A62, 58 (1944).
[86] E. Bergstrand, *Arkiv. Physik* 8, 457 (1954).
[87] S. W. Hawking, in *A Brief History of Time* (Bantam, New York, 1988), p. 81.
[88] S. W. Hawking, in *A Brief History of Time* (Bantam, New York, 1988), p. 99.
[89] R. D. Sard, in *Relativistic Mechanics* (W. A. Benjamin, New York, 1970), p. 320.
[90] W. Rindler, in *Essential Relativity*, Springer-Verlag, New York, 1977, p. 21.
[91] R. V. Pound and J. L. Snider, *Phys. Rev.* 140 (3B), B788 (1965).
[92] A. Pais, *'Subtle is the Lord...' The Science and Life of Albert Einstein* (Oxford University Press, Oxford, 1982), pp. 194-200.
[93] I. I. Shapiro, *Phys. Rev.* Letters 13, 789 (1964).
[94] A. Pais, *'Subtle is the Lord...' The Science and Life of Albert Einstein* (Oxford University Press, Oxford, 1982), pp. 178-182.
[95] R. J. Buenker, *J. Chem. Phys.* (Russia) 22 (10), 124 (2003); see also http://arxiv.org, physics/0411110.
[96] A Sommerfeld, *Atomic Structure and Spectral Lines*, 3rd English edition, translated by H. L. Brose, (Menthuen, London, 1934).
[97] R. D. Sard, in *Relativistic Mechanics* (W. A. Benjamin, New York, 1970), p. 195.
[98] W Rindler, in *Essential Relativity* (Springer Verlag, New York, 1977), p. 10.
[99] H A Lorentz, in *The Einstein Theory of Relativity*,(Brentano's, New York, 1920).
[100] C. M. Will, in *General Relativity*, S. Hawking and W. Israel, Eds., (Cambridge University Press, New York, 1979), p. 24.
[101] A Pais, in *'Subtle is the Lord ...'*, (Oxford University Press, Oxford, 1982), p. 179.
[102] S Weinberg, in *Gravitation and Cosmology*, (Wiley, New York, 1972), p. 198.
[103] R H Dicke, in *The Theoretical Significance of Experimental Relativity*, (Gordon and Breach, New York, 1964), p. 27.
[104] W Rindler, in *Essential Relativity*, (Springer Verlag, New York, 1977), p. 145.
[105] R. J. Buenker, G. Golubkov, M. Golubkov, I. Karpov and M. Manzheliy, "Relativity Laws for the Variation of Rates of Clocks Moving in Free Space and GPS Positioning Errors Caused by Space-weather Events," in *Global Navigation Systems* (ISBN 980-307-843-9; InTech-China, 2013), p. 1.
[106] T. Van Flandern, "What the Global Positioning System Tells Us about Relativity," in *Open Questions in Relativistic Physics*, F. Selleri (ed.) (*Apeiron*, Montreal, *Apeiron*; 1998), pp. 81-90.
[107] C. M. Will, in *Was Einstein Right?: Putting General Relativity to the Test*, 2nd ed. (Basic Books Inc., New York, 1993), p. 272.
[108] V. A. Fock, *J. Exptl. Theoret. Phys.* (U.S.S.R) 38, 108 (1960).
[109] R. D. Sard, in *Relativistic Mechanics* (W. A. Benjamin, New York, 1970), p. 310.
[110] W. Rindler, in *Essential Relativity* (Springer-Verlag, New York, 1977), p. 15.
[111] R. D. Sard, in *Relativistic Mechanics* (W. A. Benjamin, New York, 1970), p. 319.
[112] I. Newton, *Opticks* (Dover Publications, New York, 1952), pp. 270- 273.
[113] G. Shortley and D. Williams, *Elements of Physics*, 3rd Edition (Prentice- Hall Inc., Englewood Cliffs, N. J., 1961), p. 573.
[114] G. Peacock, Ed., *Miscellaneous Works of the Late Thomas Young* (John Murray, London,1855).
[115] A. Einstein, *Ann. Physik* 17, 132 (1905).

[116] A. H. Compton, *Phys. Rev.* 21, 715 (1923).
[117] G. N. Lewis, *Nature* 118, 874 (1926).
[118] R. P. Feynman, R. B. Leighton and M. Sands, in *The Feynman Lectures on Physics*, (Addison-Wesley Publishing Company, Reading, MA, 1963), pp. 26-27.
[119] J. Stark, *Phys. Zeitschr.* 10, 902 (1909).
[120] A. Pais, in *'Subtle is the Lord…' The Science and Life of Albert Einstein* (Oxford University Press, Oxford, 1982), p. 409.
[121] M. Planck, *Ann. Physik* 4, 553 (1901).
[122] R. D. Sard, in *Relativistic Mechanics* (W. A. Benjamin, New York, 1970), p. 156.
[123] Lord Rayleigh, in *Scientific Papers*, Vol. 1, 1881, p. 537.
[124] P. L. Muiño and R. J. Buenker, *J. Chem. Phys.* (Russia) 23 (2), 111 (2004); see also http:/arxiv.org, physics/0502100
[125] M. Born, *Z. Physik* 38, 803 (1926).
[126] H. Paul, in *Photonen: Experimente und ihre Deutung* (Vieweg, Braunschweig, 1985), pp. 98-111.
[127] I. Shapiro, M. E. Ash, R. P. Ingalls, W. B. Smith, D. B. Campbell, R. B. Dyce, R. F. Juergen and G. Pettengill, *Phys. Rev.* Lett. 26, 1132 (1971).
[128] A. Pais, in *'Subtle is the Lord…' The Science and Life of Albert Einstein* (Oxford University Press, Oxford, 1982), p. 146.
[129] G. Feinberg, *Phys. Rev.* 159, 1089 (1967).
[130] R. J. Buenker, *J. Chem. Phys.* (Russia) 23 (7), 80 (2004); see also http://arxiv.org, physics/0411109.
[131] L. J. Wang, A. Kuzmich and A. Dogariu, *Nature* 406, 277 (2000).
[132] R. J. Buenker, *Physics Essays* 26, 494 (2013).
[133] R. D. Sard, in *Relativistic Mechanics* (W. A. Benjamin, New York, 1970), p. 140.
[134] R. D. Sard, in *Relativistic Mechanics* (W. A. Benjamin, New York, 1970), p. 310.
[135] R. J. Buenker, *Mol. Phys.* **77**, 1095 (1992); see also http://arxiv.org, physics/0504070.
[136] L. Marder, *Time and the Space-Traveller* (Allen & Unwin, London, 1971).
[137] R. A. Serway and R. J. Beichner, in *Physics for Scientists and Engineers. 5th ed.* (Harcourt, Orlando, 1999), p. 1262.
[138] R. T. Weidner and R. L. Sells, in *Elementary Modern Physics* (Allyn and Bacon, Boston, 1962), p. 410.
[139] J. R. Reitz and F. J. Milford, in *Foundations of Electromagnetic Theory* (Addison-Wesley, Reading, Massachusetts, 1960), p. 23.
[140] J. R. Reitz and F. J. Milford, in *Foundations of Electromagnetic Theory* (Addison-Wesley, Reading, Massachusetts, 1960), p. 152.
[141] J. R. Reitz and F. J. Milford, in *Foundations of Electromagnetic Theory* (Addison-Wesley, Reading, Massachusetts, 1960), p. 368.
[142] R. D. Sard, in *Relativistic Mechanics* (W. A. Benjamin, New York, 1970), p. 355.

Index